THE JOY OF
BOTANICAL DRAWING

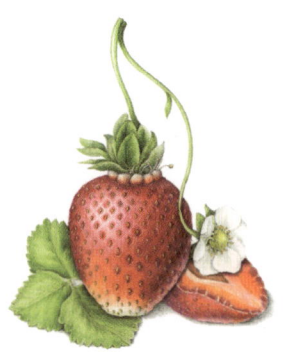

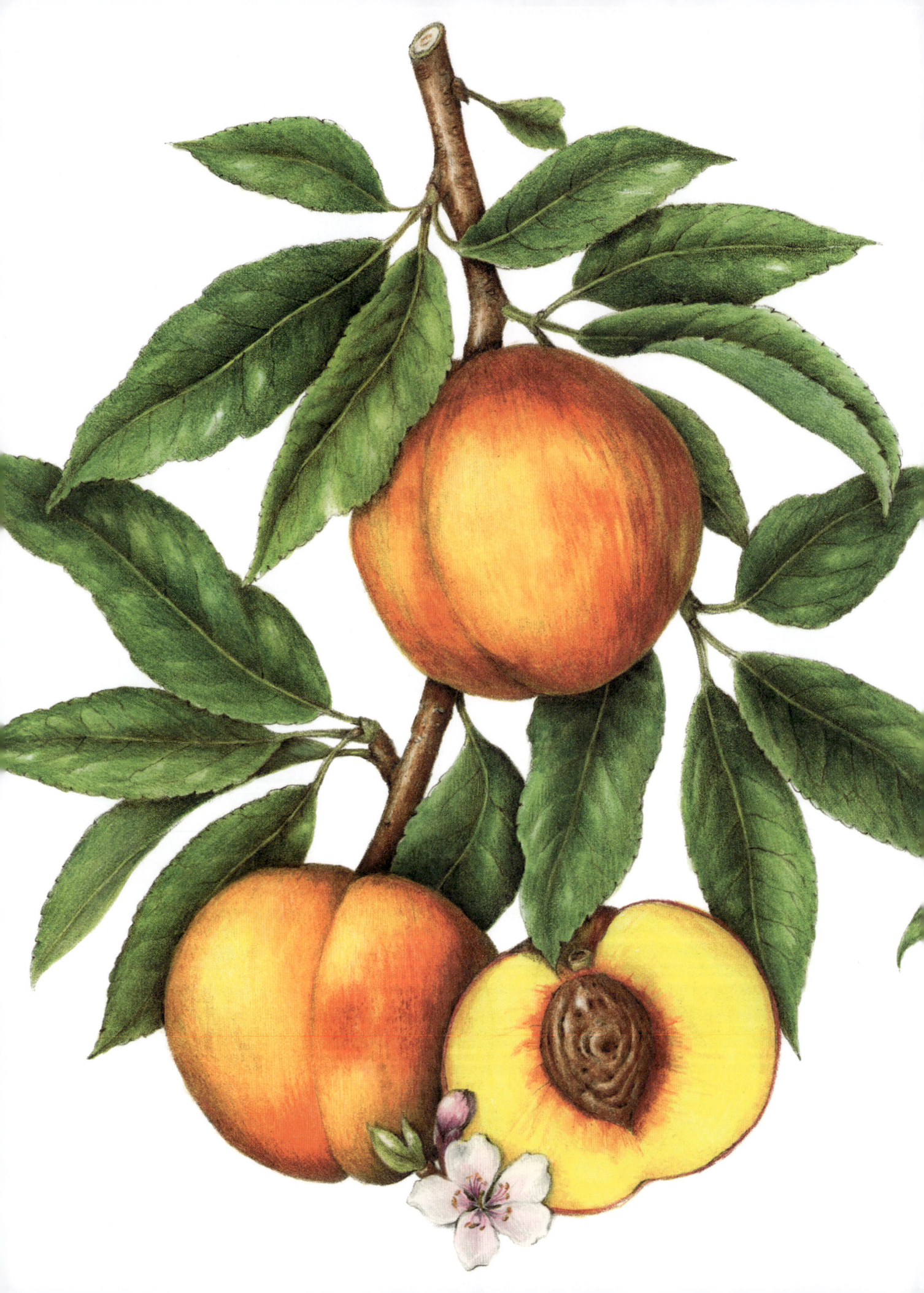

THE JOY OF
BOTANICAL
DRAWING

A Step-by-Step Guide to
Drawing and Painting Flowers,
Leaves, Fruit, and More

WENDY HOLLENDER

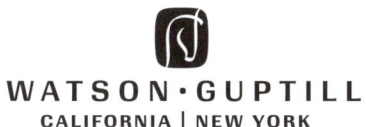

WATSON·GUPTILL
CALIFORNIA | NEW YORK

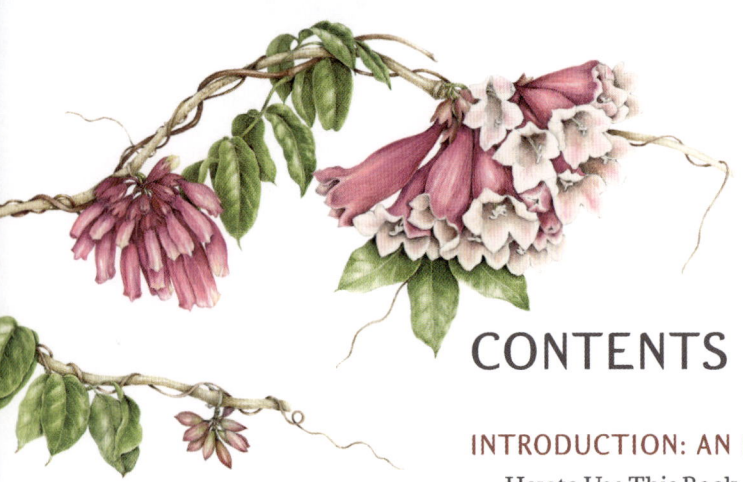

CONTENTS

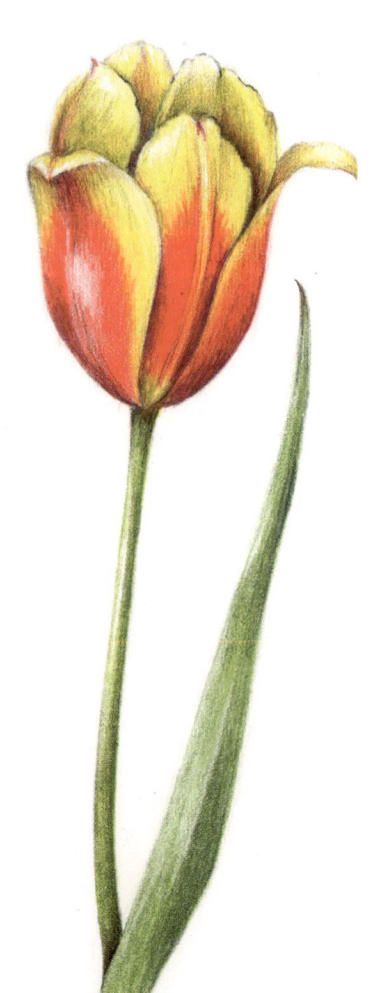

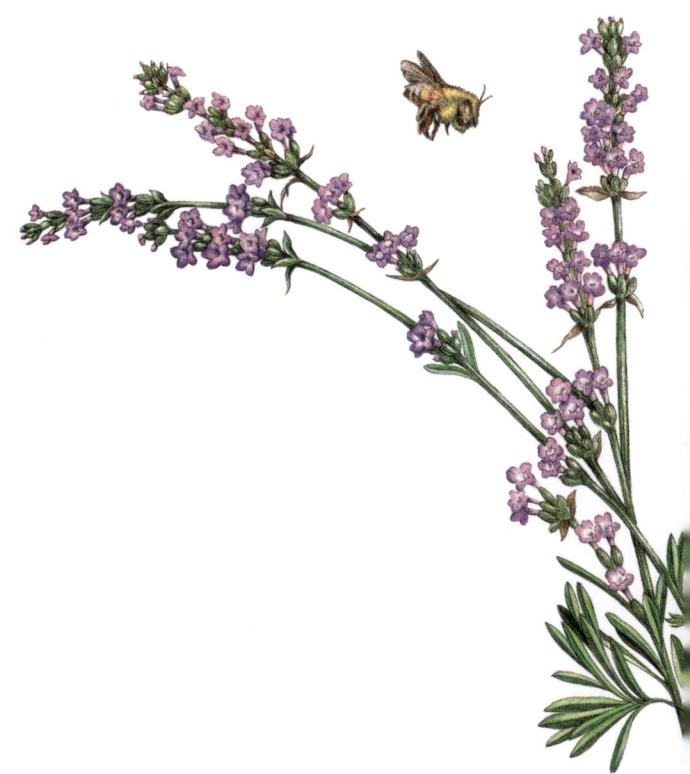

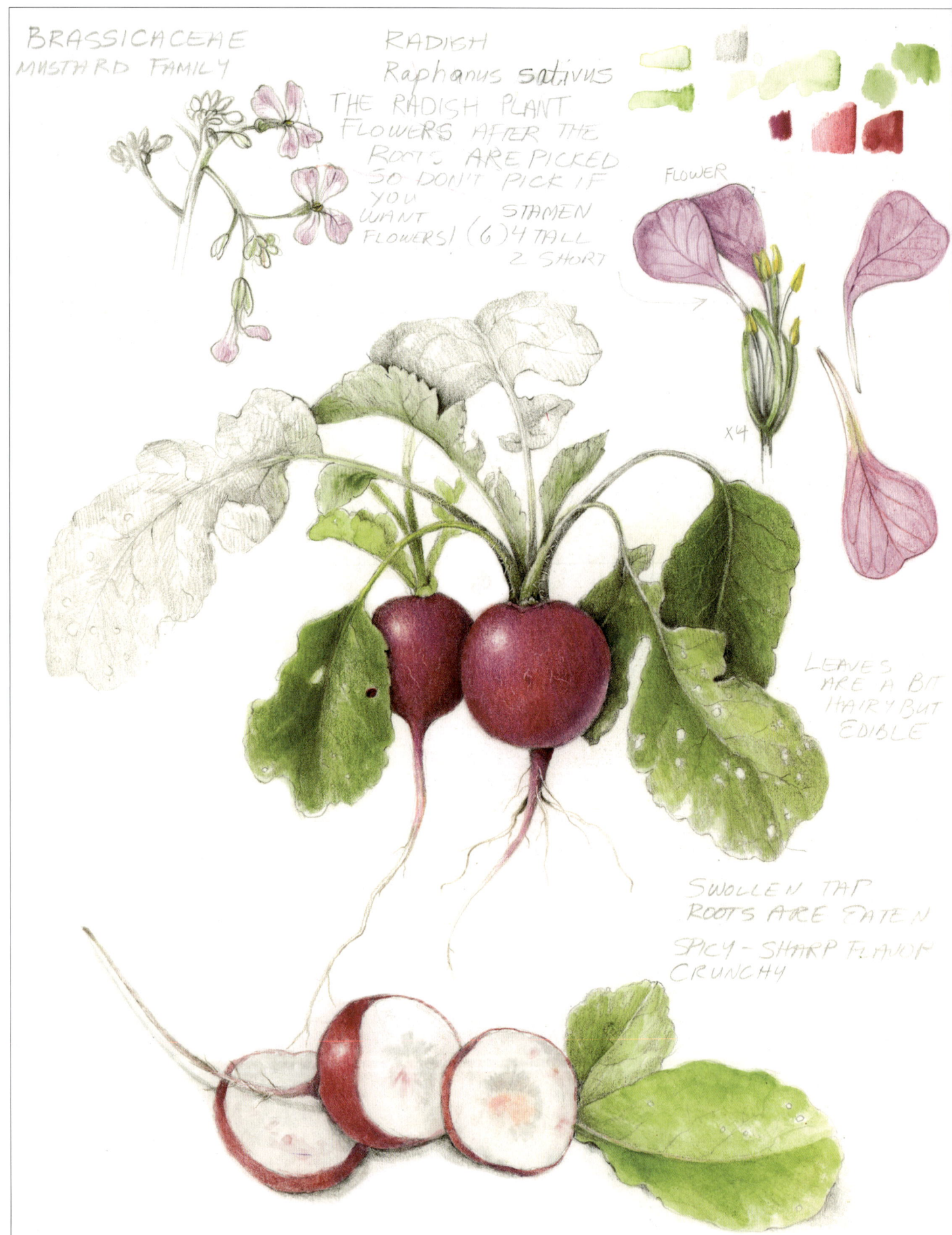

BRASSICACEAE
MUSTARD FAMILY

RADISH
Raphanus sativus
THE RADISH PLANT
FLOWERS AFTER THE
ROOTS ARE PICKED
SO DON'T PICK IF
YOU
WANT   STAMEN
FLOWERS! (6) 4 TALL
                   2 SHORT

FLOWER

X4

LEAVES
ARE A BIT
HAIRY BUT
EDIBLE

SWOLLEN TAP
ROOTS ARE EATEN

SPICY - SHARP FLAVOR
CRUNCHY

INTRODUCTION:
AN INVITATION

"The best time to plant a tree was twenty years ago.
The second-best time is now."

—CHINESE PROVERB

I first learned botanical-illustration techniques twenty years ago. The moment I understood these techniques, a door opened for me, and I immediately fell in love with the practice of botanical drawing. Since that day, it feels like the plants are leading me along a path that I steadily follow.

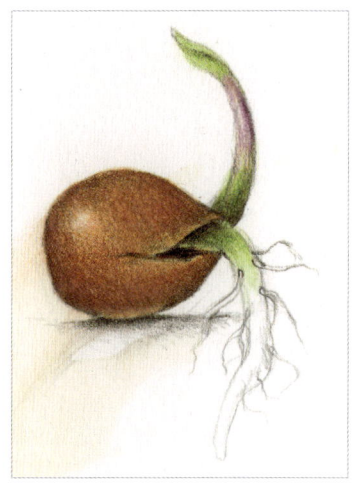

When I study and draw from nature, I feel a presence of something that never fails to take my breath away. I "undress" the plant to study the mystery within, exploring plants and flowers on a micro level, almost the way an insect does. Many people desire a deep connection with nature. They think botanical drawing is something they cannot do because they have no talent. Most of these people have never studied drawing.

I want to share two comments I have heard thousands of times from people when I tell them I am a botanical artist and that I teach this subject. The first thing they say is, "I can't draw." I ask them how they know this; have they ever studied or practiced drawing? They say, "I have never studied drawing because I have no talent." The second comment I hear from students in my drawing class is, "I am afraid . . ." This sentence ends with several things: afraid to ruin a drawing, to mess up the paper, to begin. Through the botanical drawing lessons in this book, it is my hope that you will experience the relaxing and meditative quality of drawing from nature, and develop a toolbox of drawing skills at the same time. If you slow down and practice these techniques, not only will you learn to draw but you will experience a close and personal relationship with nature. This often leads to a state of flow or well-being, so you will approach each drawing session with enthusiasm, purpose, and confidence.

You might think this book is about results. Tangible botanical drawings to hold in your hands, hang on the wall, and share with others; however, the focus of this book is not on the results but on developing the practice—and the process of doing is the best part. I think it's a privilege to immerse myself in nature's world, and it fills me with joy on a daily basis. I'm not sure why this is, but my intuition tells me it is a combination of the repetitive activity of slow drawing combined with the close-up examination of nature on an intimate level. It might even be the plants' qualities that help with this.

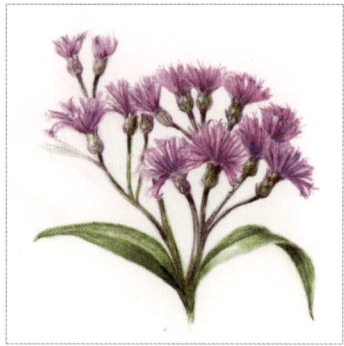

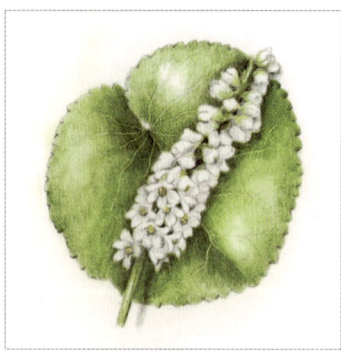

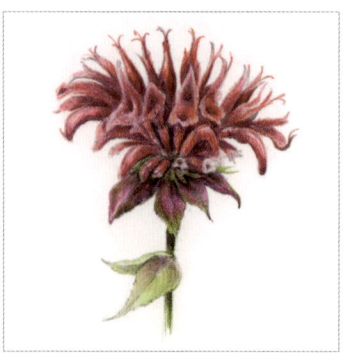

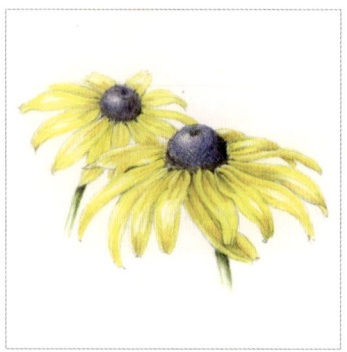

I remember once drawing a lavender flower, which has tiny flowers. While looking at the flower close up, the intoxicating scent of lavender filled the air. I kept looking closely at the flower just to get another whiff! Whatever the reason for the joy the natural world brings, the results are clear. I am happy in the doing and look forward to this every day, even at those times when I'm working on a particularly difficult subject. And I think the drawings reflect the joy of unlocking nature's complexities. The tangible result is the botanical drawing I have created, and it serves as a reminder of my time spent studying and relaxing with this plant as well as a piece of art for everyone else to enjoy and see. I am pleased to show the world the details and magic I see in the plant world, something they might not have noticed without my help.

I practiced botanical drawing happily for ten years in a large city, inspired by the urban parks and botanical gardens. As I worked in the city, the plants called to me and encouraged me to put down roots in some fertile soil of my own. I moved to a horse farm in a small agricultural community, and with the help of family and friends, we turned it into gardens full of plants for food, beauty, and inspiration. I follow the plants at home and when I travel. Year after year, I track a plant that has caught my attention so that I can see how it develops from flower to fruit. Color is usually the first thing I see, a bright color against a backdrop of green. I need to get closer, and eventually when I have studied the flower parts under magnification, I feel as if a window into nature's life cycle has opened. This has taken me to magical places all over the world where I can draw exotic, unusual, and common plants. I feel lucky that this is my daily practice.

I draw almost every day. My ideal day begins like this: I wake up in the morning, birds chirping outside my window. Coffee cup in hand, I go outside to see what nature has been up to overnight. As I walk around the gardens, I spy a new flower opening or a ripening fruit. It's always exciting to see botanical developments, whether it's a plant we have put in the garden that is thriving or an interesting weed that has shown up on its own. I am a student of nature, and I enjoy the opportunity to observe small details and remain open to nature's surprises. This feeling reminds me of a small child experiencing everything for the first time, making new discoveries and revisiting something from the day before to rejoice in the memory of it. As I walk through the gardens, something inevitably grabs my attention, and I know I want to capture it in a drawing. Once I've chosen something I'd like to draw, my day has focus and meaning. I grab my clippers and take my subject to the table where my art supply kit awaits. I sit down, observe my plant closely, and begin to draw.

I invite you to join me on this journey and develop your own process for drawing botanical subjects inspired by nature.

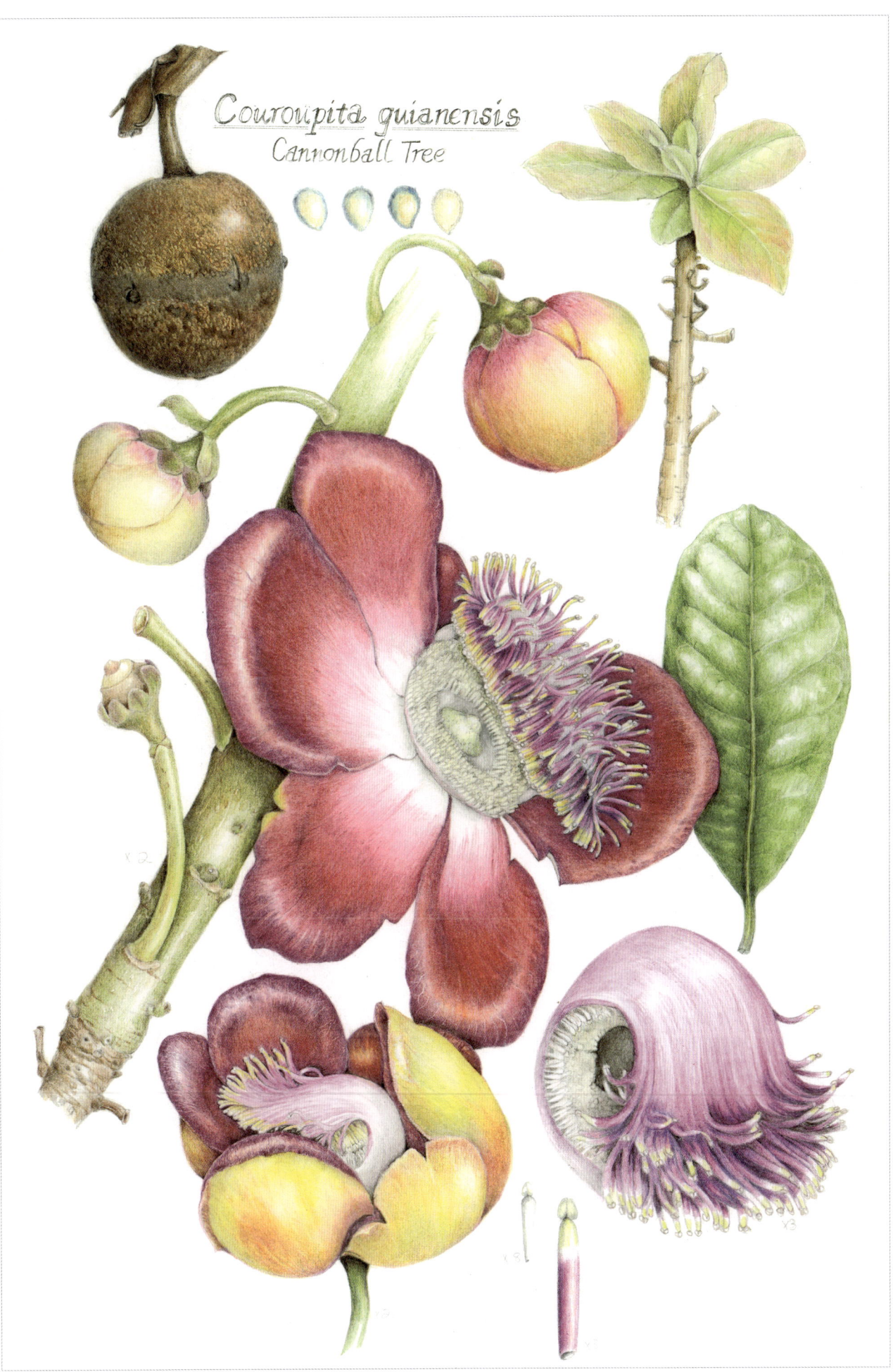

Couroupita guianensis
Cannonball Tree

How to Use This Book

This book is filled with drawing exercises. Step-by-step, I will reveal the secrets of drawing leaves, cones, fruit, plants, and flowers at a relaxed pace that will give you the opportunity to absorb the skills and develop the confidence to draw. Observing nature's quiet beauty, form, color, and life cycle encourages a systematic approach to drawing. These series of techniques can be practiced over and over again.

I encourage you to use this book like a favorite cookbook. We don't usually read cookbooks cover to cover, but by subject. For example, if you are planning on cooking a leafy green vegetable, you go to the chapter on vegetables to find instruction and a recipe. In this book, once you've practiced and learned the basics in chapters 1 through 3, feel free to drop in on any chapter to focus on various subjects and techniques. Whether you start with a lesson on a leaf or a flower, you'll still need to learn all the techniques, and there is no reason to have to draw a leaf before a flower or a branch. Think of the basics as core skills; once you have them down, then you can build on them by doing any of the lessons in each chapter. Let nature be your guide, pick your subject or let a subject pick you, and dive into the appropriate chapter or chapters! Have fun on your journey. Repeat lessons, just as you cook a favorite recipe over and over again.

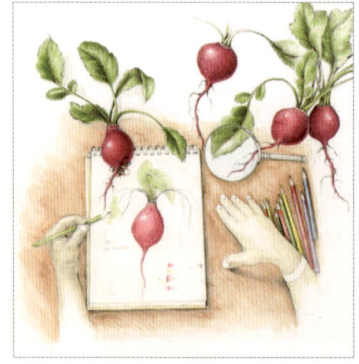

And just as you cook with real food, please make sure to always work from your own live three-dimensional subject. Use my step-by-step drawings to understand how I created the forms, but create your own original work, don't copy mine or work from a photograph. This is important, because my goal is to show you how to create the illusion of three-dimensional form on a two-dimensional surface, which requires you to translate the three-dimensional world to a flat piece of paper.

How much time will you need to devote to this practice? How long is each lesson? Everyone works at their own pace, and some of you will have more experience, so will perhaps more quickly absorb these techniques. Lessons in this book vary in the amount of time needed, but they're organized to be completed in one to three hours for smaller lessons. As you improve and tackle more complex drawings, the timing can increase. I've streamlined my techniques into short, bite-size lessons so that you can learn quickly and thoroughly. Long lessons can often lead to feelings of being overwhelmed or inadequate—or even to incompletion. I want you to feel good every step of the way with your pacing and progress so that you will gain confidence and want to continue even if you have only small bits of time to practice.

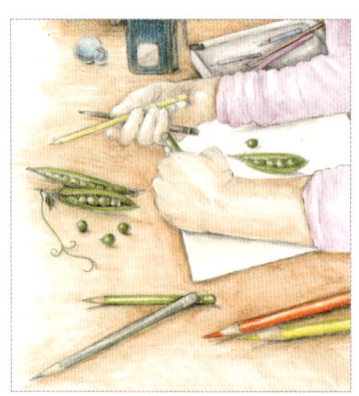

Make your own realistic time commitment, then break a lesson down into appropriate segments. I think it is important to commit to at least thirty minutes per session, and it would be ideal to work at least five days a week; however, if you can devote only one session a week to drawing, that is okay, too. I think three to eight hours a week is a good beginning goal. If you want your practice to be short and manageable, stick to small-size subjects that will fit on 5 x 7-inch paper. I will warn you that this process can be addictive, and you may find you want to work much more than this! So advise your family and friends of your newfound habit! Fortunately, this can be a social activity that you practice in the company of others and take with you on your travels.

Botanical Drawing Basics

Some people write and draw daily in sketchbooks to document nature in quick sketches. Other botanical artists work on a single painting for weeks or months at a time. I like to pick one small development in nature, perhaps a fallen nut, a drying leaf, or an open flower, and draw that one small thing—slowly—with lots of detail, structure, and realistic use of light and shadow. I follow nature, along with the changing seasons, practicing and improving my botanical drawing skills, and I learn something new each day about the plants I am drawing. Once you've learned the basics, you'll find your own motivation to continue developing your unique practice and style of drawing.

I often use colored pencil in combination with watercolor, but because the majority of layers are applied using colored pencil, I simply refer to this technique as drawing rather than painting. I believe that the repetitive act of slow drawing, building layers of tonal variation, and the gentle stroking of the pencil along the paper help me relax. This methodical process allows me to feel in control and results in a realistic three-dimensional drawing. It's possible to achieve most of these effects without using watercolor, but I encourage you to give watercolor a try because it speeds up the process, covering the white texture of the paper, and adds vibrancy to a drawing. I rely on watercolor and burnishing with colored pencil to blend and saturate color, creating a smooth surface.

Successful botanical drawings all display the following key elements: the use of a range of values with a consistent light source to describe light and shadows on form, perspective, and plant structure, as well as the use of realistic color. And I'll discuss these techniques in this book. The *grisaille* method, which I call "grisaille toning," uses neutral toning to describe form first and color second. Once I create the form in a neutral tone using a single primary light source, I add color to the tonal drawing with a layer of watercolor wash. When you combine form and color this way, you create a solid foundation as you build layers of color and value. It's easy to become seduced by colors at the beginning of a drawing and unintentionally create flat forms, so the grisaille technique helps to avoid this. The first series of colored pencil and watercolor wash exercises in this book will guide you as you draw from real plants and turn simple geometric forms into shapes in nature: a cylinder into a branch, a sphere into a tomato. As your core skills improve, you will draw using perspective and study the plant structure of leaves and flowers. These lessons will ultimately familiarize you with all the characteristics of a complete plant, in all stages of development so that you can draw a detailed botanical illustration that tells a particular plant's story from roots to fruit.

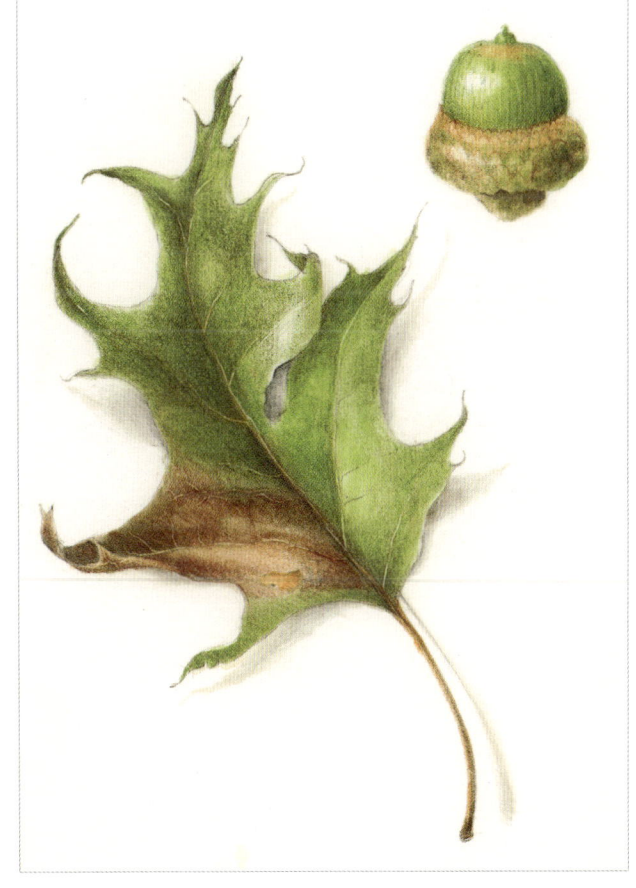

The Daily Practice

I am always on the lookout for subjects to draw. I collect nuts, seedpods, leaves, and whatever else interests me. A small area is always set up with my drawing supplies ready and subjects to draw. I also enjoy listening to music, books, and podcasts while I work and have those ready to go as well. This ensures that I am more likely to make time in my day for drawing.

In warmer weather, I set up an outdoor space for working so that I can enjoy the light outside, a summer breeze, and the outdoor sounds of nature as I draw. Using a concise amount of art supplies allows for a simple work space that doesn't require much setup or cleanup. I often work on small paper, which is quite portable to take with me wherever I go.

Follow these steps to set up your daily practice:

· Choose an inviting place to work with good light.

· Set up your art supplies where you like to draw and a travel case ready to pack for drawing on-the-go.

· Collect subjects and have them out and waiting. They will be calling you to the drawing table!

· Schedule your drawing time when you know you can fit it in.

What are you waiting for? Now is the perfect time to start!

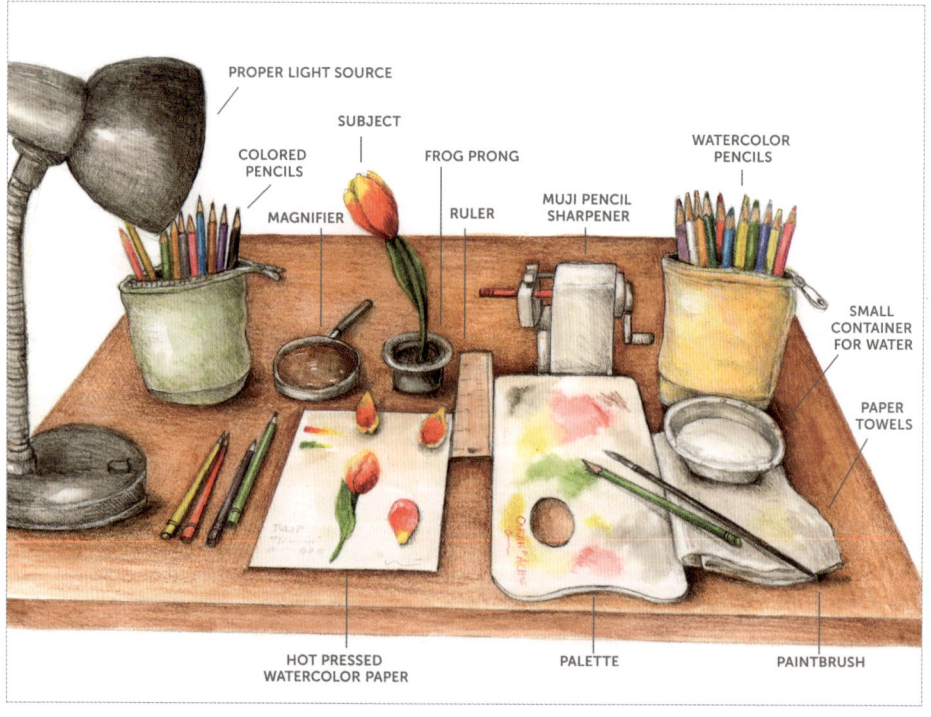

PROPER LIGHT SOURCE

SUBJECT

COLORED PENCILS

FROG PRONG

WATERCOLOR PENCILS

MAGNIFIER

RULER

MUJI PENCIL SHARPENER

SMALL CONTAINER FOR WATER

PAPER TOWELS

HOT PRESSED WATERCOLOR PAPER

PALETTE

PAINTBRUSH

Materials

I am specific about the art supplies that I suggest here because I have "test driven" an array of materials over the years and have streamlined what I now use and recommend. Experience has shown me which pencils work the best, the minimum amount of colors that I need, the best papers that work for my techniques, and other supplies that are needed regularly. You can start with whatever you have on hand, but most everyone ends up wanting some of what I recommend—and usually all of it! If you're on a tight budget, stick with the essential pencil colors to start, and consider the student grade. Because most art supply stores do not carry individual pencil colors, I started an online store years ago. This way you can buy all the supplies I recommend in one place. My website is drawbotanical.com, and I have art supply kits that include the basics and also a kit that has everything you'll need from soup to nuts. I also recommend online resources such as dickblick.com.

Colored Pencils

I recommend Faber-Castell's oil-based Polychromos line, which I will refer to as "colored pencils"; they sharpen to a very sharp point, and the application texture is very fine on hot press paper. These pencils allow for a slow buildup of many transparent layers, and the gradual development of a complete range of values and colors can result in a drawing that looks extremely realistic. These colored pencils also erase a bit, which is helpful, so I can worry less when I work, knowing I can erase carefully if needed. All Faber-Castell pencils have superior lightfast ratings, and the colors available are excellent.

Faber-Castell's Albrecht Dürer pencils, which I also use and call "watercolor pencils" throughout the book, are available in the same colors as the Polychromos line. This makes learning my favorite colors and mixing both kinds of pencils a bit easier. Note that the pencils look similar but the shape of the pencil shaft is different, and they are marked with the appropriate name. Polychromos are round and Albrecht Dürer pencils have a hexagon shape.

Keep your pencils separate, so as not to use the much softer watercolor pencil when you want to build fine layers of tone with colored pencil.

FABER-CASTELL POLYCHROMOS COLORED PENCILS

I recommend a list of twenty-five colors to start. If you want to gather supplies slowly, be sure to get the eleven starred (∗) colors below and at right, and wait on the others. I don't recommend sets of pencils, because there are often colors in them that you may never use—especially the greens. In botanical drawing, green is so important, and it's a waste of time to use the greens that come in the smaller sets of pencils because they are not realistic to nature. If the green is too bright, you end up spending lots of time adjusting the color and it is so much better to just start with the greens that are closer to nature's colors.

Primary and Secondary Colors

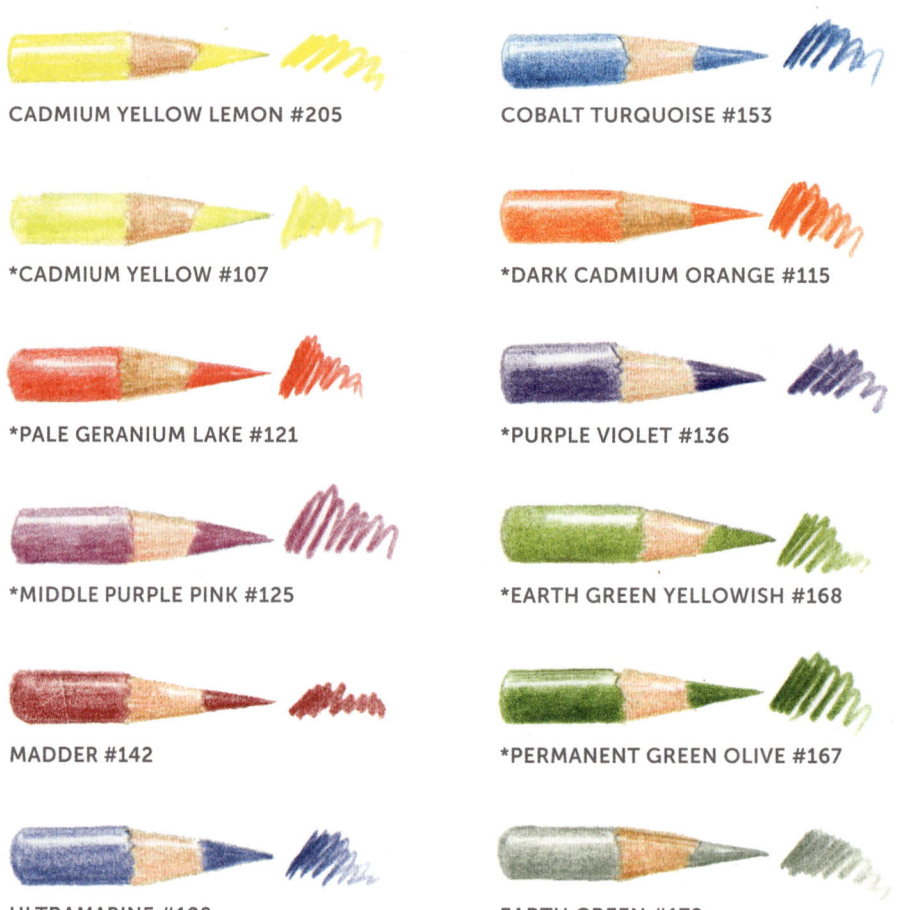

CADMIUM YELLOW LEMON #205

*CADMIUM YELLOW #107

*PALE GERANIUM LAKE #121

*MIDDLE PURPLE PINK #125

MADDER #142

ULTRAMARINE #120

COBALT TURQUOISE #153

*DARK CADMIUM ORANGE #115

*PURPLE VIOLET #136

*EARTH GREEN YELLOWISH #168

*PERMANENT GREEN OLIVE #167

EARTH GREEN #172

Dark Colors to Mix for Deep Shades of Color

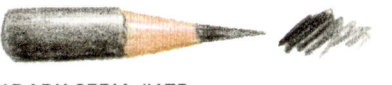

*DARK SEPIA #175

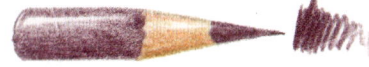

*RED VIOLET #194

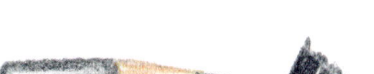

DARK INDIGO #157

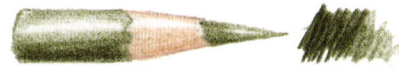

OLIVE GREEN YELLOWISH #173

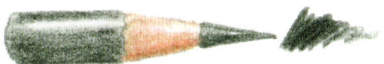

CHROME OXIDE GREEN #278

Light Colors for Tints, Highlights, and Burnishing

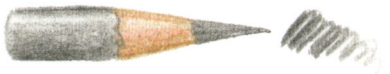

WARM GREY IV #273

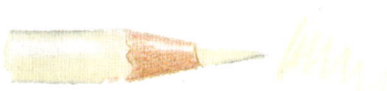

*IVORY #103

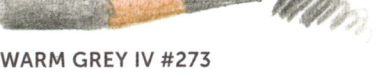

WHITE #101

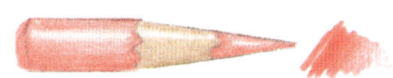

DARK FLESH #130

EARTH TONES

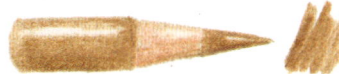

BURNT OCHRE #187

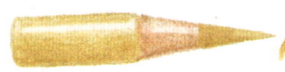

YELLOW OCHRE #183

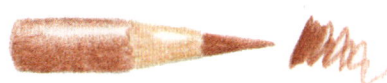

VENETIAN RED #190

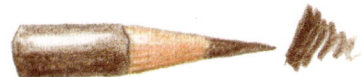

*BURNT SIENNA #283

Watercolor Pencils

Utilizing the basic colors in watercolor pencils allows you to combine watercolor and colored pencil together in a drawing. I find this technique enables you to work faster and create vibrant, colorful botanicals, yet retain a good amount of control. The detail work is done with colored pencil, which is an easier medium to control than watercolor alone. My favorite watercolor pencils, as noted previously, are also from Faber-Castell, called Albrecht Dürer. They have vibrant colors that match the Polychromos colored pencil set and an excellent lightfast rating. Faber-Castell also offers a student-grade line called Goldfaber that is very good if you are on a tighter budget, and the color names are the same as the Polychromos line.

FABER-CASTELL ALBRECHT DÜRER WATERCOLOR PENCILS

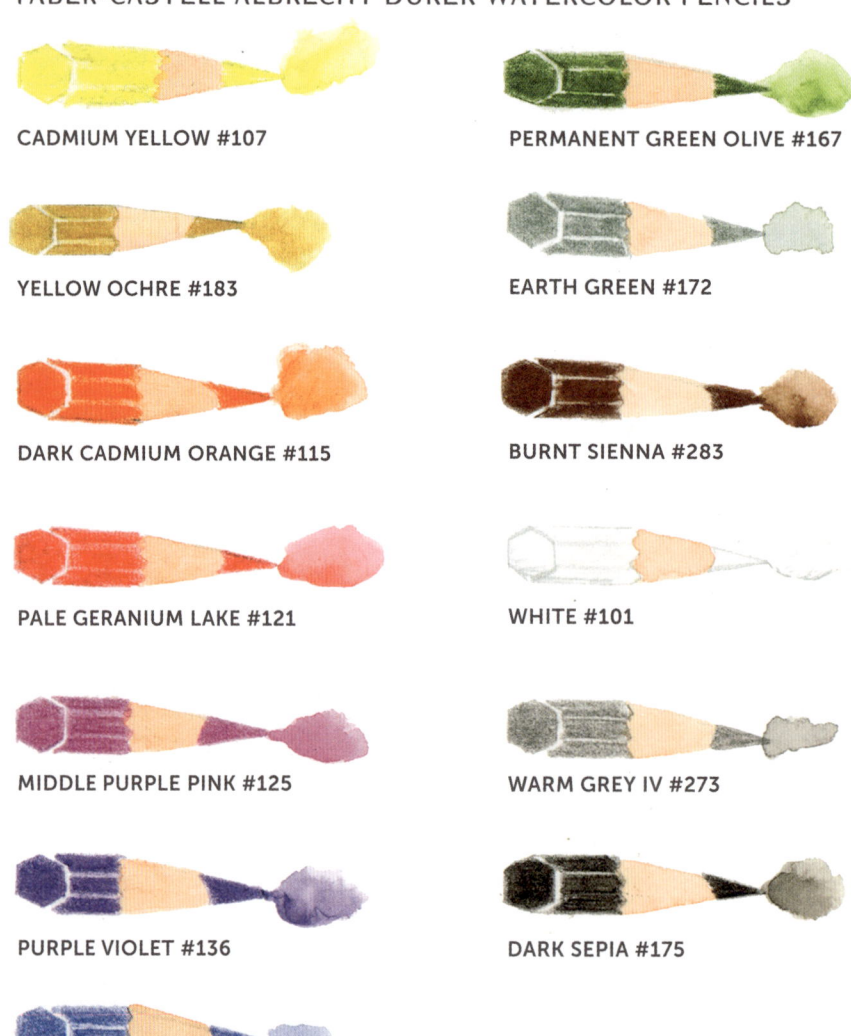

CADMIUM YELLOW #107

PERMANENT GREEN OLIVE #167

YELLOW OCHRE #183

EARTH GREEN #172

DARK CADMIUM ORANGE #115

BURNT SIENNA #283

PALE GERANIUM LAKE #121

WHITE #101

MIDDLE PURPLE PINK #125

WARM GREY IV #273

PURPLE VIOLET #136

DARK SEPIA #175

ULTRAMARINE #120

Additional Pencils

You'll also need one or two graphite pencils with H lead, and I like Tombow brand pencils for this. An H pencil is neither too dark nor too hard. It works well for doing a light drawing with precise measurements.

At the end of a drawing, I like to use Prismacolor Verithin pencils in Dark Brown, Black, and Cool Grey 70%, which have a harder point for creating clean sharp edges, fine details, and subtle cast shadows. Sometimes I even like to burnish with these pencils, as the fine point gets inside the texture of the paper and fills it in easily.

Pencil Sharpeners

I can't stress enough the importance of a good pencil sharpener. We'll be working with a very sharp point most of the time, and a good sharpener will give you a consistent, sharp point without having to work too hard to get it. My favorite kind is a hand-cranked option such as the Muji Desktop Pencil Sharpener or a Carl Angel-5 desktop sharpener. The Muji sharpeners are available in Muji shops and on the company website, as well as on my website; and the Carl Angel-5 can be ordered on Amazon.

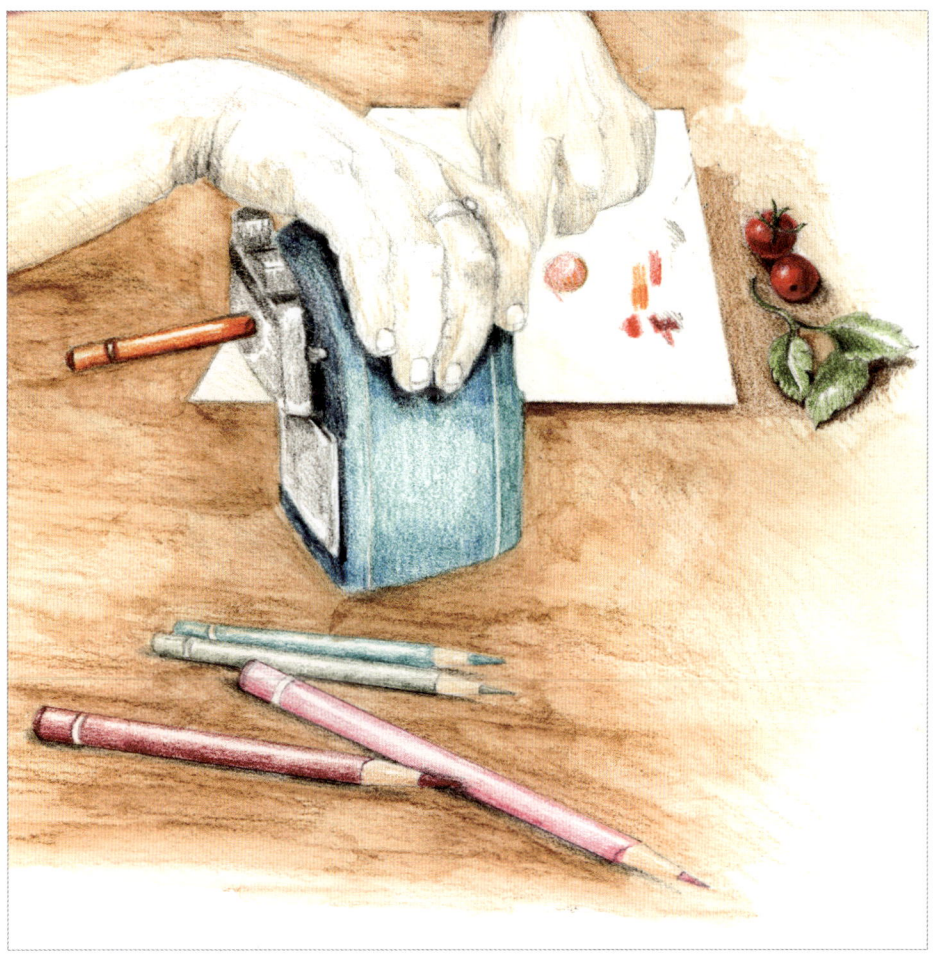

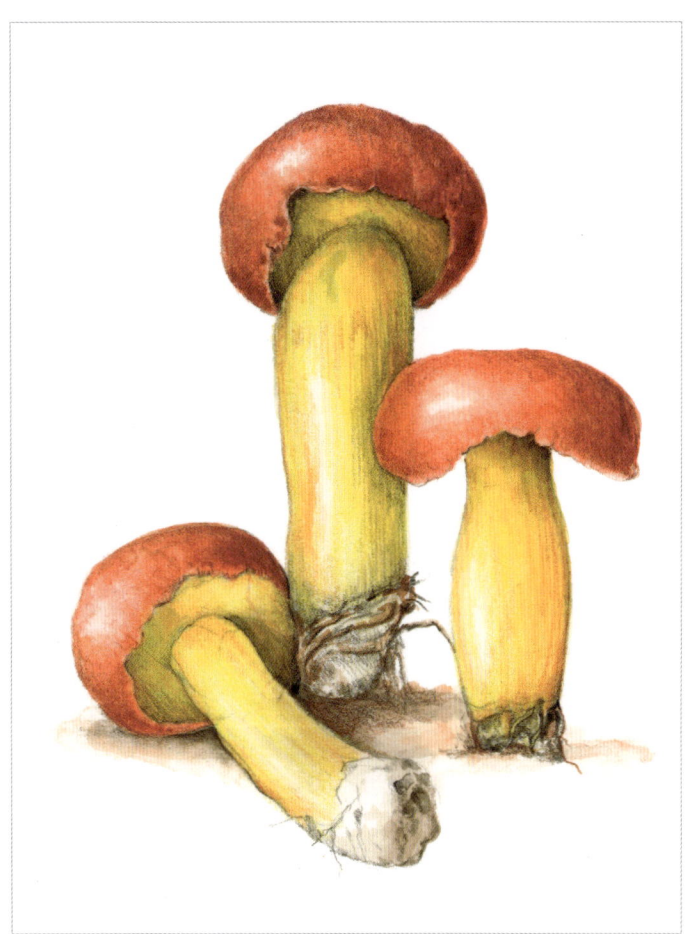

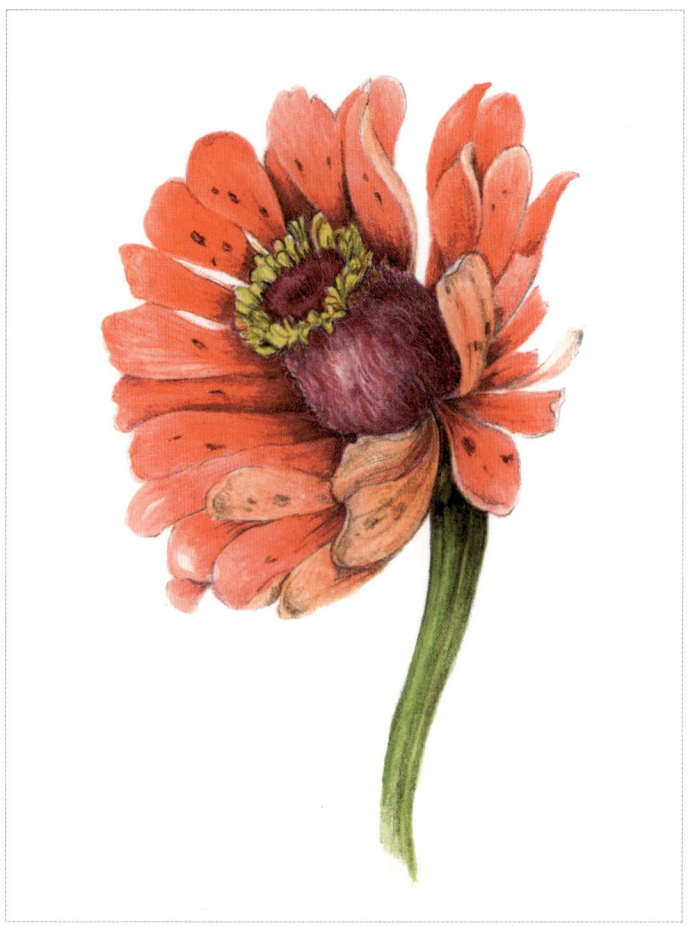

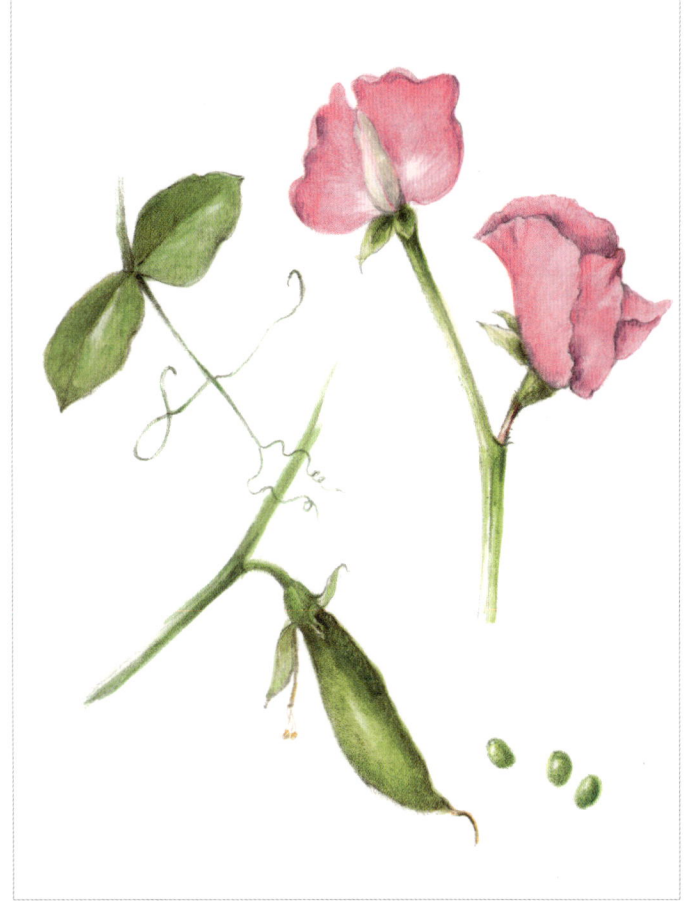

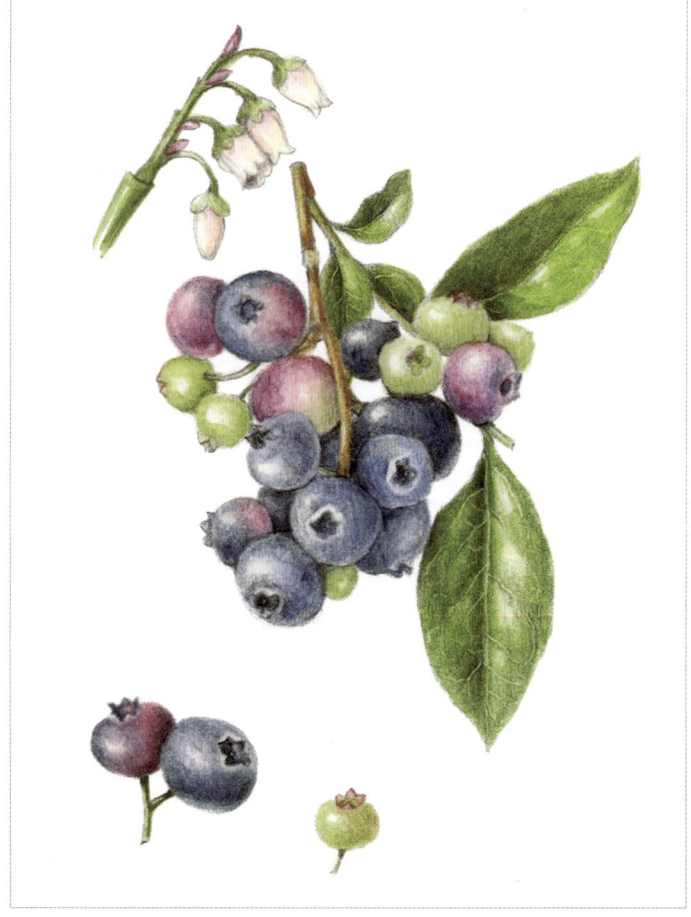

Erasers

Because these Polychromos pencils erase, having erasers for different uses is important. For light erasing I use a kneaded eraser, and the Tombow Mono Round Zero is great for fine, precision erasing. When you have to erase a large or dark area, the colored pencil eraser is great, just don't erase too hard or you might create a hole in the paper.

Brushes

You'll need a few watercolor brushes in two or three sizes. My favorites are synthetic brushes from the Japanese company Interlon, and I use sizes 3/0 (small), 2 (medium), and 6 (large). I like them because they are inexpensive and work well, and I don't have to worry about keeping an expensive brush in good shape. I just use them until they no longer have a good point, and then I start a fresh one. I use three sizes so that I can control the amount of watercolor I apply. For dry-brush techniques and fine details, I use the small brush. For small forms, such as stems, I use a small- or medium-size brush. For larger subjects, such as a big leaf, I will use a size 6 brush.

Many watercolor botanical artists prefer sable brushes, and a popular line is the Winsor & Newton Series 7 Kolinsky Sable Pointed Round, which is expensive. The techniques using watercolor in this book rely mostly on washes (not fine, dry brushwork), so you won't necessarily need top-quality brushes, especially in the beginning. When on the go, the Pentel Aquash Water Brush is convenient, as there's a place to store water inside the brush.

Paper

I recommend using a small 5 x 7-inch spiral pad of hot pressed watercolor paper for continuity. This size paper is big enough to capture intimate subjects, yet small enough to allow for smaller compositions that can be completed in relatively short time periods. A smaller size paper is much less intimidating and relieves the pressure of wasting a good piece of paper.

Hot press watercolor paper, which is smooth, is my favorite to use for combining colored pencil and watercolor. I do not like a rough, cold press watercolor paper, as it dulls your pencils and doesn't allow for a fine line and buildup of fine texture. I like to work on a spiral pad of paper, but they aren't available in the United States with hot pressed watercolor paper, so I custom make them in two sizes using Stonehenge Aqua paper and a spiral binding that can be opened and closed to add and remove paper. Blocks of paper or sheets cut up into smaller pages are fine, too. Brands that I like are Stonehenge Aqua by Legion, Fabriano Artistico, Arches, and the Stillman & Birn Zeta Series.

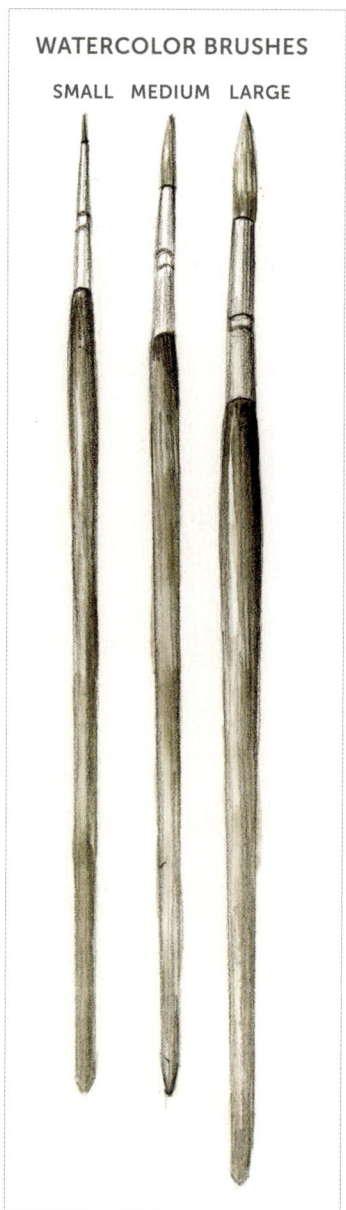

WATERCOLOR BRUSHES

SMALL MEDIUM LARGE

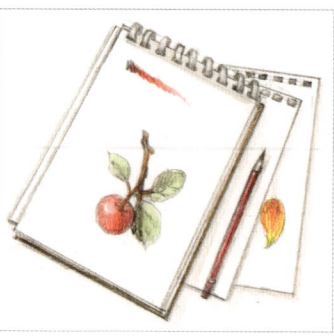

Other Materials

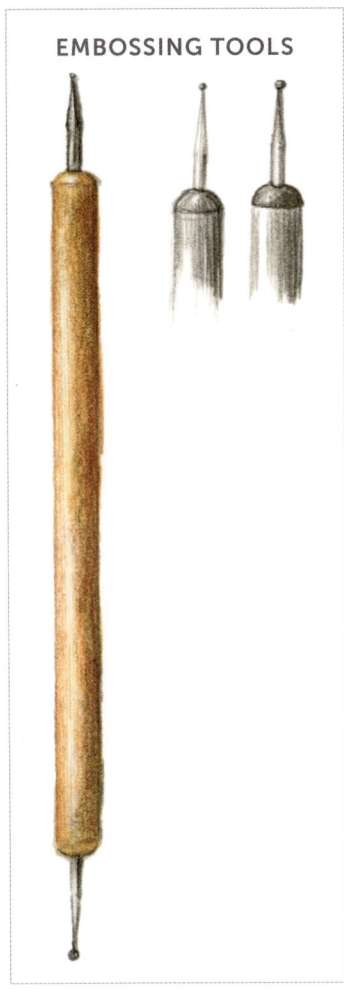

EMBOSSING TOOLS

WESTCOTT 6" CLEAR RULER OR OTHER SMALL CLEAR RULER: A small ruler is essential for measuring your subject accurately and quickly. It will save you lots of time in the end.

COLLAPSIBLE WATER CUP BY FABER-CASTELL: You will need a small container to hold water while you work.

EMBOSSING TOOLS IN TWO OR THREE SIZES WITH SMALL ROUND ENDS: Embossing tools help with retaining fine lines, hairs, and dots on a drawing. A set of embossing tools with small round ends works great for this purpose. Sets are available in art and crafting supply stores. I carry them in my online supply shop. I don't recommend one particular brand because they are always changing, and they all work well.

WAX PAPER: Small sheets of the kind used in the kitchen to help with embossing techniques.

TRACING PAPER: Any tracing paper will do, and it doesn't have to be expensive. Used to do quick sketches, to create cross-contour lines over a drawing to help with understanding a three-dimensional surface, and to plan compositions.

MAGNIFYING GLASS (4X) OR JEWELER'S LOUPE (30X): To closely inspect subjects.

DRAFTSMAN MINI DUSTER OR OTHER BRUSH: For wiping away debris.

CARAN D'ACHE PALETTE AQUARELLE OR GRAFIX DURA-LAR FILM: For creating watercolor washes with watercolor pencils.

PORTABLE LAMP: Any kind will do.

FROG-PRONG FLOWER HOLDER: To hold subjects (optional).

X-ACTO KNIFE: To dissect flowers and subjects.

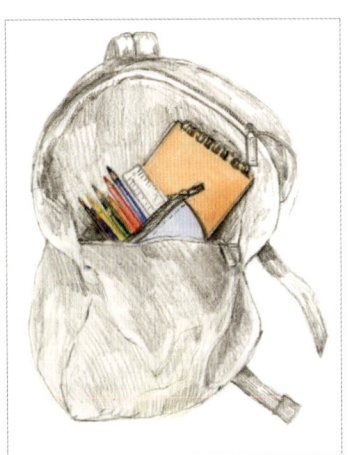

BASIC TOOL KIT

Materials you'll need for every lesson, in addition to the supplies listed in each lesson:

- Colored and watercolor pencil sets
- Dusting brush
- Erasers
- H graphite pencil
- Hot press watercolor paper
- Palette
- Pencil sharpener
- Portable lamp
- Ruler
- Scrap piece of any paper (to put under your hand as you work so that you don't smudge your drawing)
- Small water container
- Two paper towels, folded up

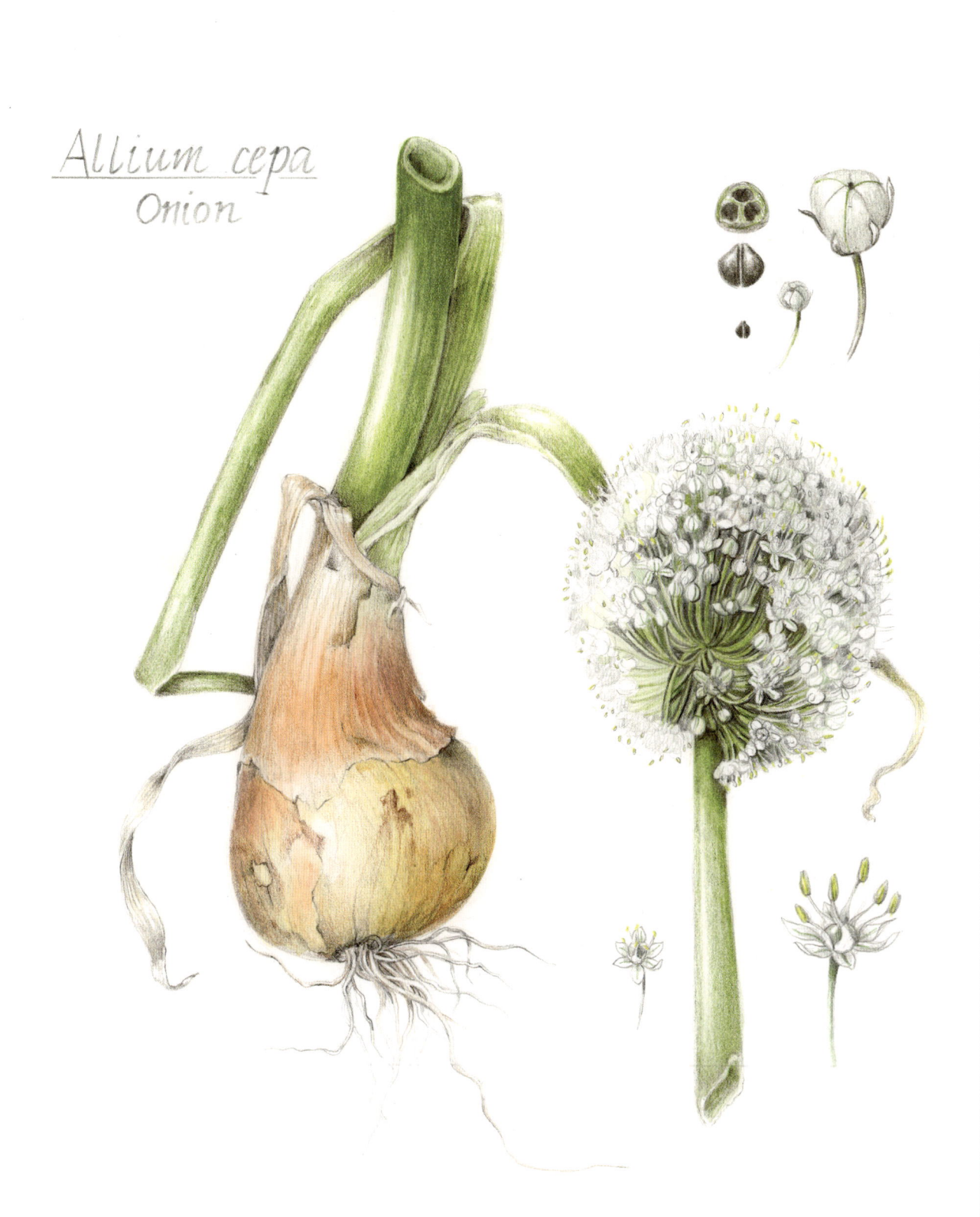

Allium cepa
Onion

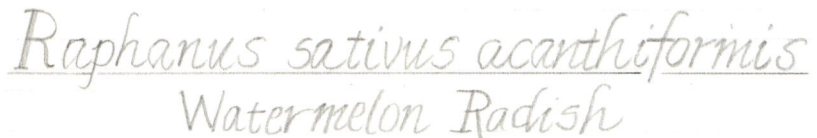

Raphanus sativus acanthiformis
Watermelon Radish

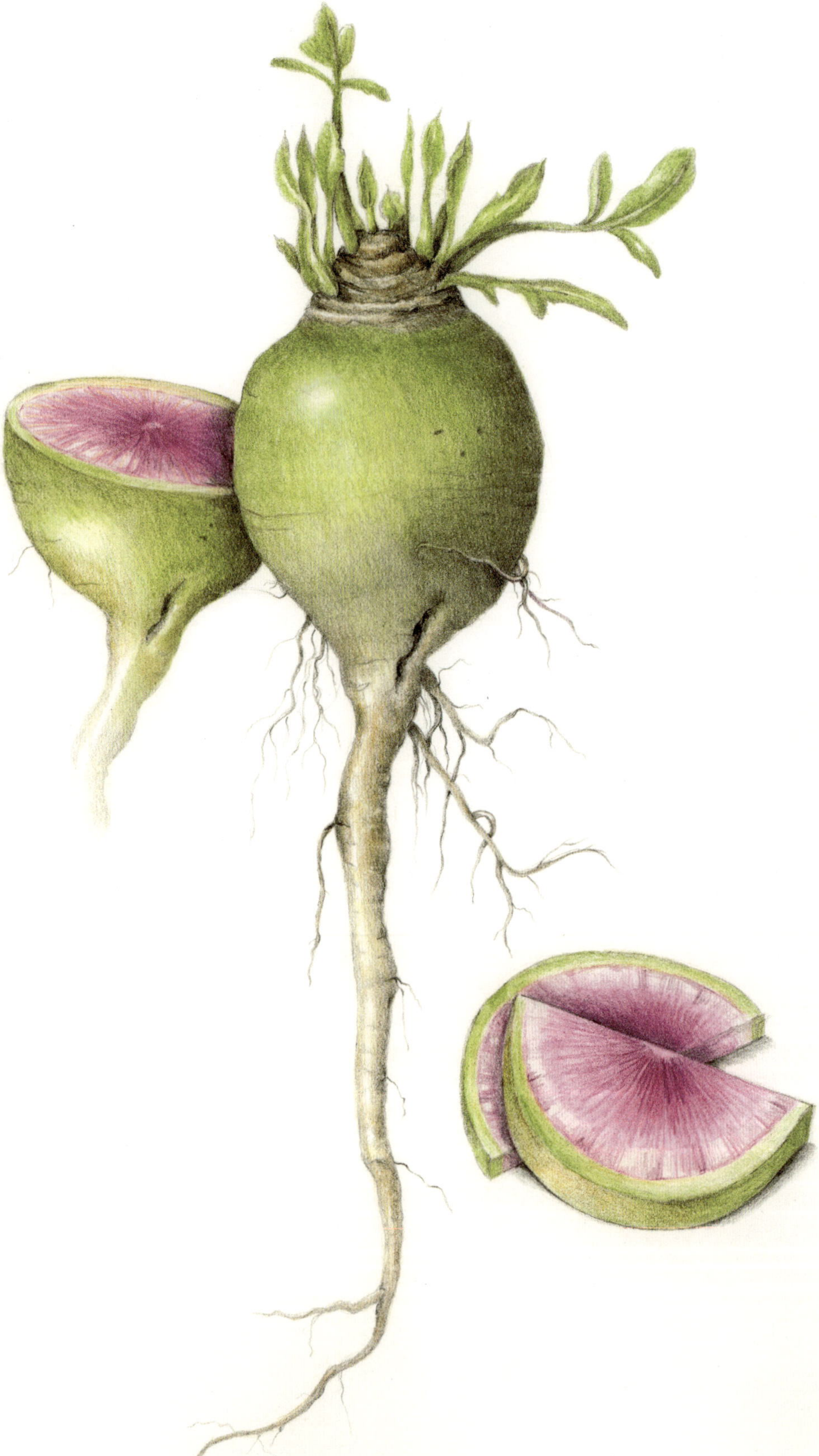

CHAPTER ONE

UNDERSTANDING LIGHT SOURCE TO CREATE THREE-DIMENSIONAL FORM

Creating a detailed, three-dimensional plant drawing feels like magic to me. Perhaps this is because I tried for a long time to make my drawings look this way and failed. Something was not quite right. It might have been the perspective, the proper proportions, or a consistent use of light and shadow. I didn't have a clear focus on how to accomplish what I wanted. I proceeded in a haphazard way, sometimes creating an area of a drawing that worked but never knowing how I achieved the success, and I was unable to do it consistently.

One of the most important concepts in botanical drawing is to understand how to use a single, specific light source to describe three-dimensional form. In this chapter, I will show you what I mean by a consistent light source and how to visualize it on forms in nature—even if you don't actually have a light illuminating your subject this way. We will start with simple geometric shapes, practice using this idealized light source on these forms, and then immediately explore similar complex forms in nature.

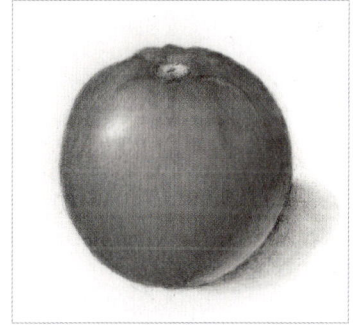

To understand how light behaves on three-dimensional forms, observe the illustration [A] of an ideal light-source setup on page 21. As you progress, you'll find that the light source isn't a mystery to solve, just a visualization exercise that will get easier and easier to do. To better understand the importance of using one light source on forms, sometimes we'll translate a photo of our subject into gray tones. Discarding the color information can help you visualize how to create shadows, midtones, and highlights on form in the proper places. Gray tones show the placement of light and shadow without the added distraction of color.

Your goal is simple at first: make your form look three-dimensional with one consistent light source. This is your mantra. Tell yourself this over and over again, because I promise you, at first you will forget and go back to the way most of us approach drawing. We tend to think we have to draw the shadows and highlights exactly as we see them, even if we don't have a good light source set up. But soon you will know where you want to put the highlights and shadows, even if they aren't visible on your subject.

When you are drawing, use a light that illuminates your subject exactly as described in "Creating a Consistent Light-Source Setup" on page 20; eventually you will learn to keep this ideal light in your head even if you don't have a light on your subject the way you will be drawing it. We will call this the imagined "head light." Your drawing is not a crime scene for which you need to document every detail as you see it but rather a drawing that will convey your subject either in a light source that you create in your head or in a light source that is actually illuminating your subject. This means that if there are multiple highlights on your subject because light is hitting it from many locations, you will choose the best highlight to show and not draw them all. Extremely harsh shadows will also be toned down so as not to overpower the drawing. Learn how simple geometric forms receive this light source. Once you memorize the light on the simplified form, you can translate the same head light onto more complex forms in nature. We'll start with a sphere, a cylinder, a cup, a cone, and an open book as reference subjects.

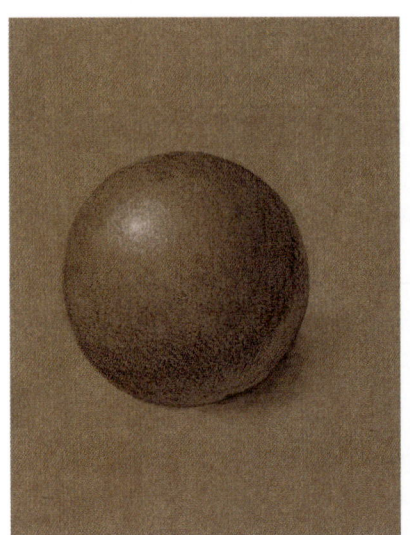
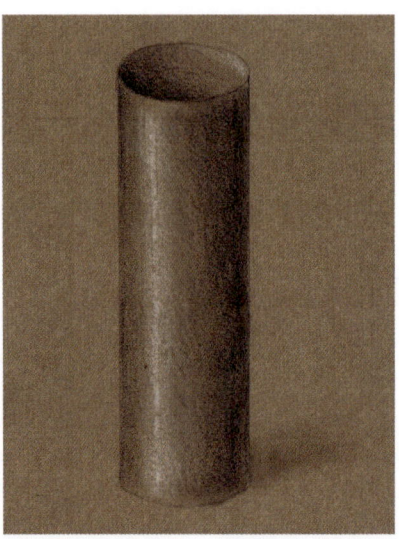
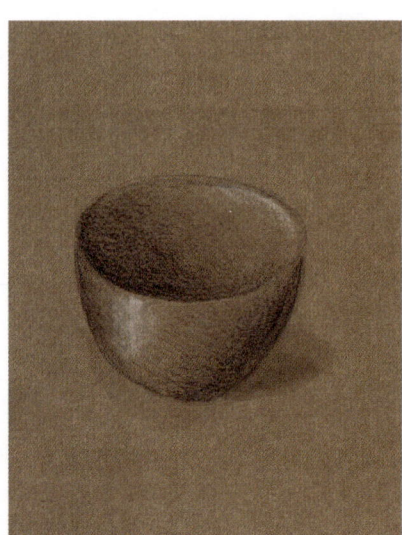
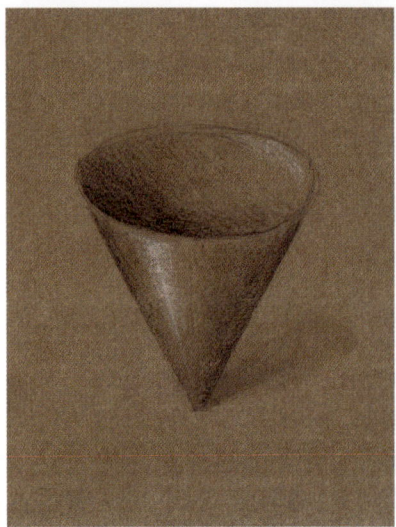
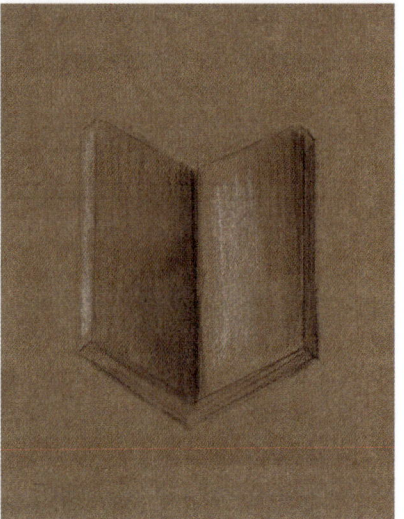

TERMS TO KEEP IN MIND

These are the terms I'll explain and use throughout the book, but I've gathered them here so that you can reference them easily if you ever forget what one means.

BURNISHING: Blending colors for a smooth finish, usually with a light-colored pencil.

CAST SHADOW: A darker shadow area that falls on the surface on which a subject is sitting. Cast shadows are created when the form of the subject blocks the light that is hitting it. A shadow is not a form itself, so it should be extremely subtle. It should be darkest next to the edge of the shape that is casting the shadow. Shadows graduate from dark to very light and should not end in a sharp line but should fade away into the color of the paper.

CONSISTENT LIGHT SOURCE (HEAD LIGHT): Your setup or imagined light source is always placed on the left or right, in front of your subject. The light hits the subject at a 45-degree angle, and so approximately one-third of your subject is in shadow while two-thirds is illuminated.

CONTOUR LINES: An outline that follows the edges or shape of a form.

CROSS-CONTOUR SKETCH: Cross-contour lines travel across a form to help describe the three-dimensional surface and how it bends away from and toward light. These quick sketches can help you visualize the placement of light and shadow. They also often follow the venation pattern (arrangement of veins on a leaf) on a form.

HIGHLIGHT: The bright areas on the surface of a form (usually on the highest points of a three-dimensional form) created by the light source hitting it directly. Even if you don't see highlights on your form, you can put them in to convey this important structural message in a drawing.

LOCAL OR DOMINANT COLOR: The main color of your subject.

OVERLAPPING: The spots where forms intersect each other, defining foreground and background. For example, when a leaf rolls over or curls or when one form is slightly in front of another, it creates an overlap. Rendering overlaps properly enhances space, depth, and structure.

REFLECTIVE HIGHLIGHTS: A reflective highlight is a weak light that bounces off of a surface onto a form or subject. It helps to differentiate between two overlapping forms or separates a form from a cast shadow. The reflective highlight should not be as light in value as a highlight. It should be very subtle, at a value of 4 or 5 (see "Create a Nine-Value Tone Bar" on page 26). It's best when it feels like glowing light and is not just a straight, empty line.

PICTURE PLANE: For our purposes, think of the picture plane as an imaginary perpendicular window right up against your subject. Taking measurements on this plane helps you to measure in perspective.

THUMBNAIL SKETCH: This simple, quick, small shape drawing of your subject is a reminder of how the light illuminates your subject so that you know where to put the shadow side, the midtones, and the highlight.

TONE OR VALUE: These terms refer to how light or dark a color appears.

Creating a Consistent Light-Source Setup

Practice visualizing light on form (use your head light) to help you make decisions on how to draw your form, even when you aren't seeing a single light source. It is also important to understand that even if you don't actually see the highlight and shadows, you can still put them on the form in your drawing. It takes a while to get used to this concept, so it's best to practice and remind yourself of it. Remember your mantra: make your form look three-dimensional with one consistent light source.

Look at my two examples of citrus drawings (below). The tangelo shows light coming from the upper right, while the Meyer lemon shows light coming from the left. The effect is the same but just from opposite sides. If possible, turn off all other light sources in the room and close window shades so that your subject is receiving light from only your light-source setup. The box can help block out other light sources. Note: If you are right-handed, I recommend placing this light on the upper left. If you are left-handed, place this light on the upper right. The goal is to have a good source of light on your subject that doesn't cast a shadow on your paper from your drawing hand as you are drawing.

Position your box on a table and place the lamp in front of the box on the left side. (A simple gooseneck lamp is fine, though you can also use a specific light that is closer to daylight, such as an OttLite. On the fly, I light my subject with a flashlight or my cell phone light to help get an overview of the highlight, the midtones, the shadows, and the cast shadows.) The light should be hitting your subject at about a 45-degree angle from the upper left or right. This position is your consistent light source.

LIGHT FORM ON MODEL

RECESSED
HOLE

HIGHLIGHT WITH
SURROUNDING
GLOW

CAST
SHADOW

LOCAL COLOR
MIDTONES

SHADOW   REFLECTIVE
HIGHLIGHT

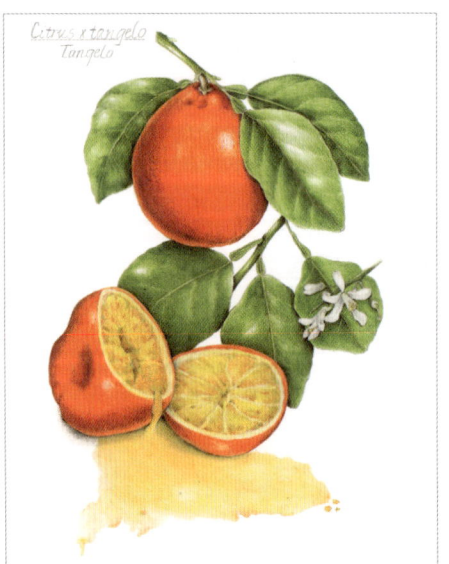

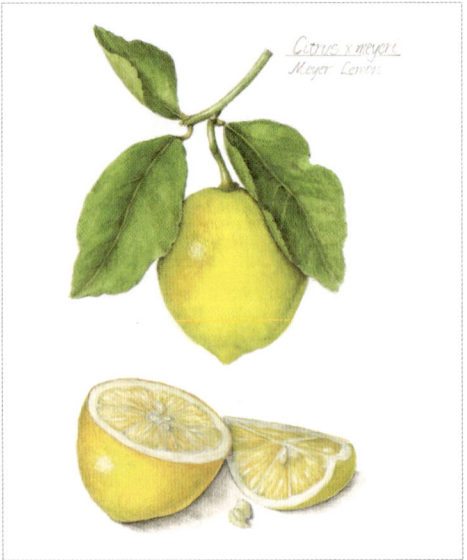

1. Place a round subject such as a tomato or an apple in the box and notice how the light hits the form. Take time to study the placement of the highlight and how the values gradually change, getting darker and darker until the darkest shadow. If you see a reflective highlight, take note of it, then pay attention to the cast shadow that is created when your subject blocks light from the surface on which it's sitting. **(A)**

2. Snap some photos with your cell phone or a camera for reference of this correct light source on your subject. In your photo-editing program, try changing the color mode to gray scale to see more clearly (without the distraction of color) how the gradual shift in values is essential to create three-dimensional form.

3. Draw a rough thumbnail sketch of the light source on your subject to help memorize where the shadows and highlights will be placed. **(B)**

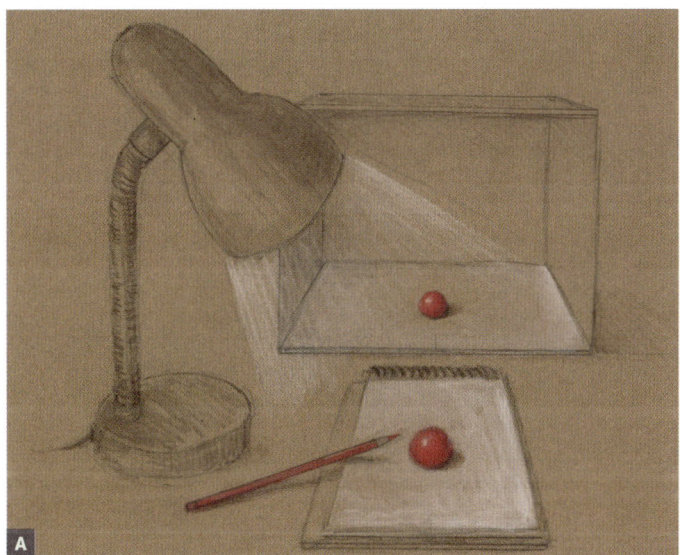

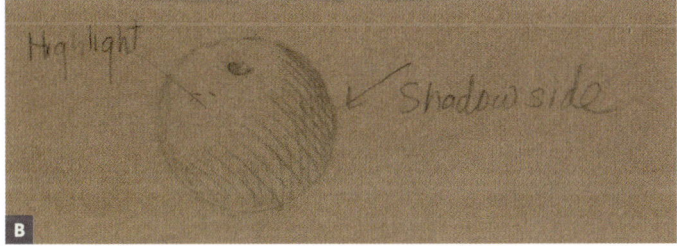

CONTINUED

4. Now move your light source and look at your subject to see how the light hits the form from different directions so that you realize this consistent light source is the best light-source model to use to make your form look descriptive. It is also good practice to go outside on a sunny day and position yourself with the light source of the sun behind you from the left and position a subject in the correct place.

5. Next, switch your subject to a cylindrical shape, such as a small branch. Follow the same procedure for looking at the form. Snap a photo and draw a thumbnail sketch. Try a cup shape and a cone shape using the same procedure, and finally practice on an open-book shape. You can create a quick model of a simple leaf out of white paper that is folded to create two distinct planes, or look at a leaf in the light to visualize how light interacts with two distinct planes. **(C)**

6. Practice looking at leaves in natural light to see the two planes created by a midvein and study leaves on plants outside. Look at other subjects: flowers, seedpods, even architectural elements, to practice visualizing light on form.

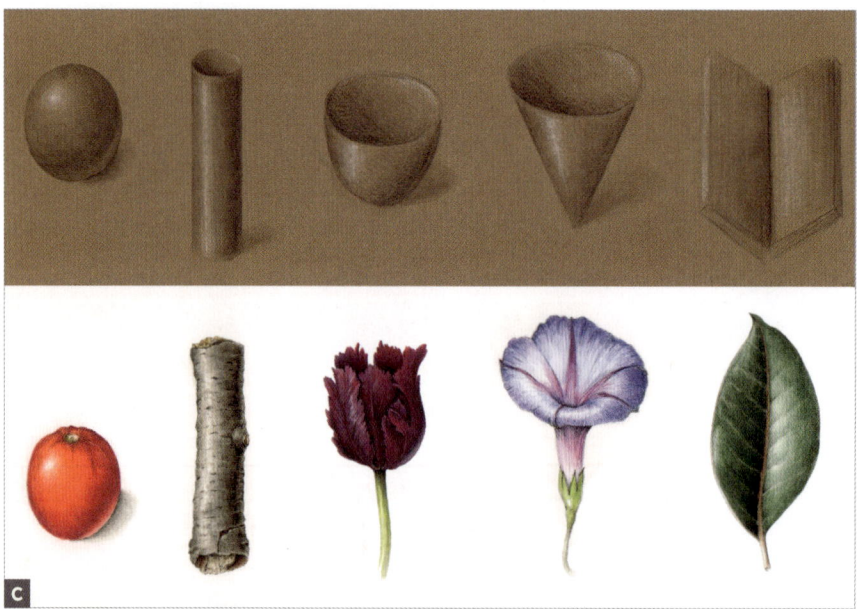

How to Use Watercolor Pencils for Botanical Drawing

There are several ways to use watercolor pencils, but this is my favorite. When you draw on a plastic palette made for watercolor pencils, or a substrate plastic, such as Grafix Dura-Lar Film or Yupo synthetic paper, the surfaces have enough friction to lay down the watercolor pencil, so when water is added to the pigment, it turns into paint. You can mix your watercolors to make other colors and also apply the watercolors directly to your drawings from the paint you mix on the palette. The palettes can also be washed and reused over and over again.

ADDITIONAL MATERIALS

• Watercolor pencils (any colors)

1. Draw a small area of pigment onto your palette.

2. Dip your favorite brush in a tiny bit of water and mix it into the watercolor on your palette. You can mix colors easily this way and control the ratio of pigment to water on your brush. This is how I recommend you work the majority of the time.

3. Paint a few swatches of color onto hot press watercolor paper, and experiment with the amount of water you add to the pigment. Paint a light color, a medium color, and a dark color. Also try swatches where you first wet a small area of paper and then apply the color wash onto the wet paper to experience how the watercolor spreads easily this way.

4. Next, draw directly on your paper with the watercolor pencil and spread the pigment with a wet brush afterward. Try to experiment with different amounts of pigment and water to see the effects this creates. This method is great to do when traveling and just making quick sketches, and it works well when using a water brush (such as a Pentel Aquash Water Brush).

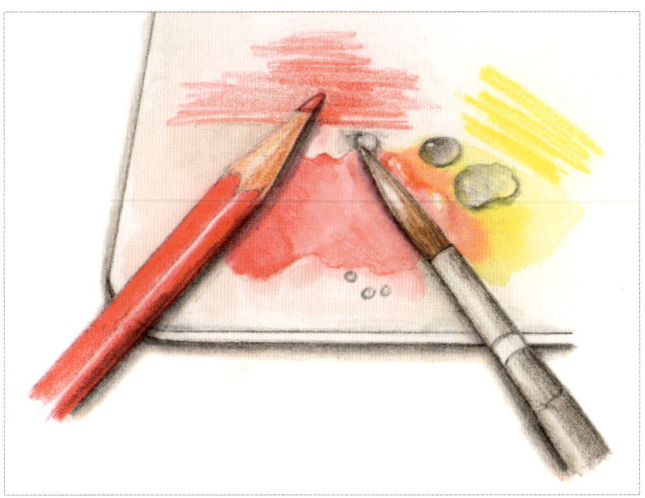

Slow-Toning to Create a Seamless Blend of Values from Light to Dark Using Watercolor and Colored Pencil

ADDITIONAL MATERIALS

- Dark Sepia watercolor pencil #175
- Watercolor brush #2
- Dark Sepia colored pencil #175

WATERCOLOR PENCILS

#175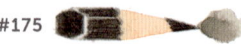

COLORED PENCILS

#175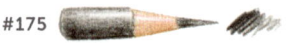

In this lesson I introduce the mechanics of creating tonal variation with a slow buildup of layers. Please focus on these two concepts as you practice the step-by-step techniques. I'll create tonal variation with a slow buildup of layers starting with a watercolor tone bar and then colored pencils, which change gradually from the lightest tone (the white color of your paper) to the darkest tone (as dark as you can get with your darkest pencil). This is the key to making forms look extremely three-dimensional. It works best to understand this concept by first working only in neutral tones without the distraction of colors.

Note: You can practice all lessons without using watercolor, but this requires more layering of colored pencils and more burnishing. It's a slower technique but allows for more control for the beginner. The watercolor speeds up the process. With these techniques, combined with the understanding of a consistent light source that you've just acquired from the first lesson, you'll start to gain confidence. At the same time, you'll feel relaxed (because repetitive, slow drawing feels good), and soon a feeling of well-being will come over you. These exercises are really fun to do when you take the time to do them slowly and methodically, and you'll find that they become an important component for all your drawing. Soon your confidence level will rise, and you will no longer be telling yourself that you have no talent and cannot draw!

Create a Tone Bar

I like to let my watercolor dry completely before drawing on top of it. How long is this? It depends on your climate and how wet you got your paper. I would say it's a good idea to wait twenty minutes for your paper to dry, but waiting even longer is always better. I often work on more than one drawing or area at a time so that while one area of watercolor is drying, I can work on something else. If you try to draw on top of your watercolor while it is not completely dry, the paper will not take the pencil and sometimes the pencil can start to chew up the paper.

1. On your watercolor pencil palette, draw an area of Dark Sepia watercolor pencil (about 1 inch in size).

2. To test the value of the Dark Sepia, dip your medium-size brush into clean water and mix it into a tiny bit of the Dark Sepia watercolor pencil. Diluted, it should be a very light value. Test the watercolor wash first by painting three small squares, one light and the other two with less water in the pigment so that you get a midtone and a dark value. **(A)** If there is too much water on your brush, blot it on your folded paper towel and then dip it back into the paint.

3. To create the tone bar, lightly draw a ½ by 3½-inch area with a graphite pencil on your paper. **(B)**

4. Paint this area with clean, plain water, starting on the right side of the tone bar and pushing the brush out to the left. Apply Dark Sepia watercolor about half to three-quarters of the way across, then blot your brush quickly on a paper towel and continue painting into the light side, making the transition to the paper seamless. **(C)**

5. Next, while this is all still wet, add a bit more Dark Sepia pigment on your brush and apply, starting on the right side and moving toward the middle in a seamless transition. Horizontal brushstrokes work well to help with this.

6. While you wait for your watercolor tone bar to dry, follow the procedure on page 26 to create a nine-value tone bar to use as a reference.

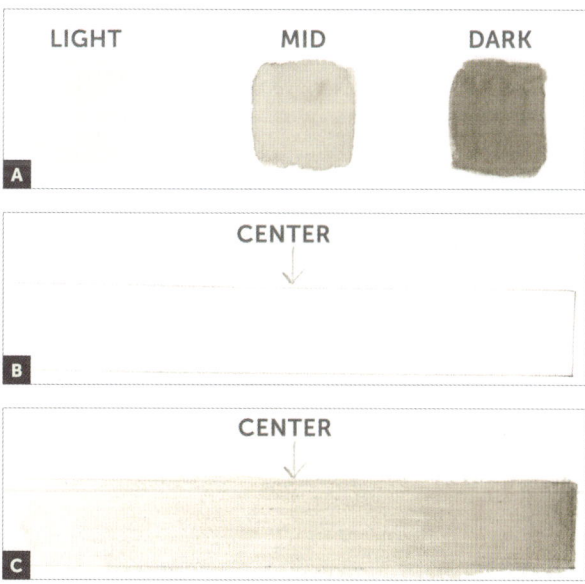

Create a Nine-Value Tone Bar

This tone bar will be used as your guide for a complete range of values from shadow to highlight. You can use this guide to help you evaluate your drawings and make sure you have created a complete range of values on each subject.

1. First draw nine small boxes underneath the watercolor tone bar, each about ⅜ inch wide.

2. Number each segment from left to right, starting with 1 on the left and ending with 9 on the right. **(A)**

3. Fill each segment with tone values from dark (9) to light (1) with your Dark Sepia colored pencil. Start on the right side and fill in a dark value, as saturated as possible, for your darkest number 9 value. Leave number 1 empty. With the side of your pencil, gently lay down a very light layer in number 2. Box number 3 might have two light layers of tone so it is just a bit darker than number 2. Number 4 is a bit darker, and number 5 has perhaps five light layers of tone.

4. Now switch to number 8 and get dark, but not as dark as number 9. Number 7 will be a tiny bit lighter than number 8, and number 6 will be darker than number 5 but lighter than number 7.

5. Make adjustments in each box so that you have nine distinct values, gradually stepping from dark to light. This is now your guide for a complete range of values from shadow to highlight. **(B)**

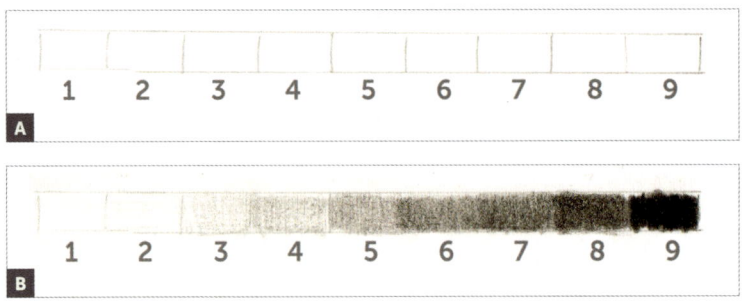

Smooth, Seamless Toning with a Neutral Colored Pencil

To create smooth, seamless toning in your tone bar, start on the right side with slow toning, then with each layer, cover a little bit less of the tone bar, leaving the left side of the tone bar lightest. Switch the directions of your strokes, and even consider turning your paper to continue laying down tone in a relaxed, comfortable manner. This is very important. You want this process to be a very slow buildup of values, and for each layer, apply smaller and smaller strokes. The final layers of strokes become tiny circular strokes that fill in the light areas of fine texture on the paper, creating a seamless, blended tone bar. If you are feeling a sense of calm and well-being, chances are you are applying your tone correctly.

You will use this technique, or develop your own variations of this technique, throughout the book. It is most important to find your own method of achieving the goal of creating many layers of seamless tonal variation. Notice the feeling of space, depth, and movement from dark to light even in a simple tone bar such as this. In the end, you should feel relaxed and proud of your tone bar. If you work too quickly and don't have a nice, smooth transition with all nine values, try another tone bar. It is fun to allow yourself the luxury of creating slow, seamless tonal drawings. And not only is it fun, but it is an essential skill to develop. Once mastered, you will be on your way to creating realistic three-dimensional drawings. Who knows, this could become your daily meditation!

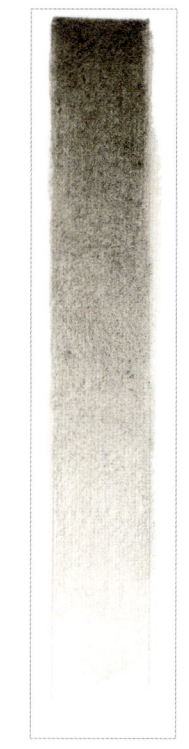

1. Make sure your watercolor wash tone bar (that you created in the lesson on page 24) is completely dry. Apply a first layer of Dark Sepia, starting on the right side, with close, vertical strokes, then make your strokes farther apart until they end at the middle of the tone bar. **(A)**

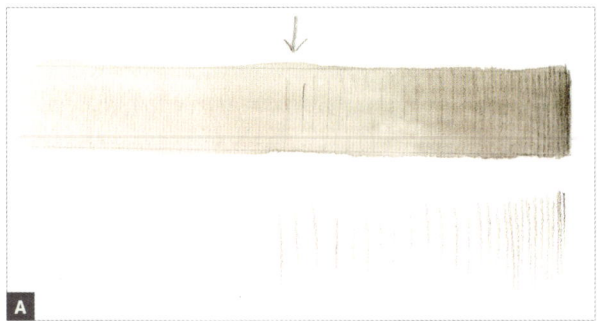

A

CONTINUED

2. Hold your pencil toward the top end, and starting again on the right, apply the next layer of tone in a horizontal direction. In order to start getting a smooth seamless transition of tones from light to dark, lay down horizontal strokes that are about ⅜ inch long and then do another group of horizontal strokes in between those, overlapping them a little bit. Keep adding horizontal strokes in this way to about the middle of the tone bar. **(B)**

3. Add more tone in a diagonal direction **(C)**, followed by another layer going in the opposite diagonal direction. **(D)**

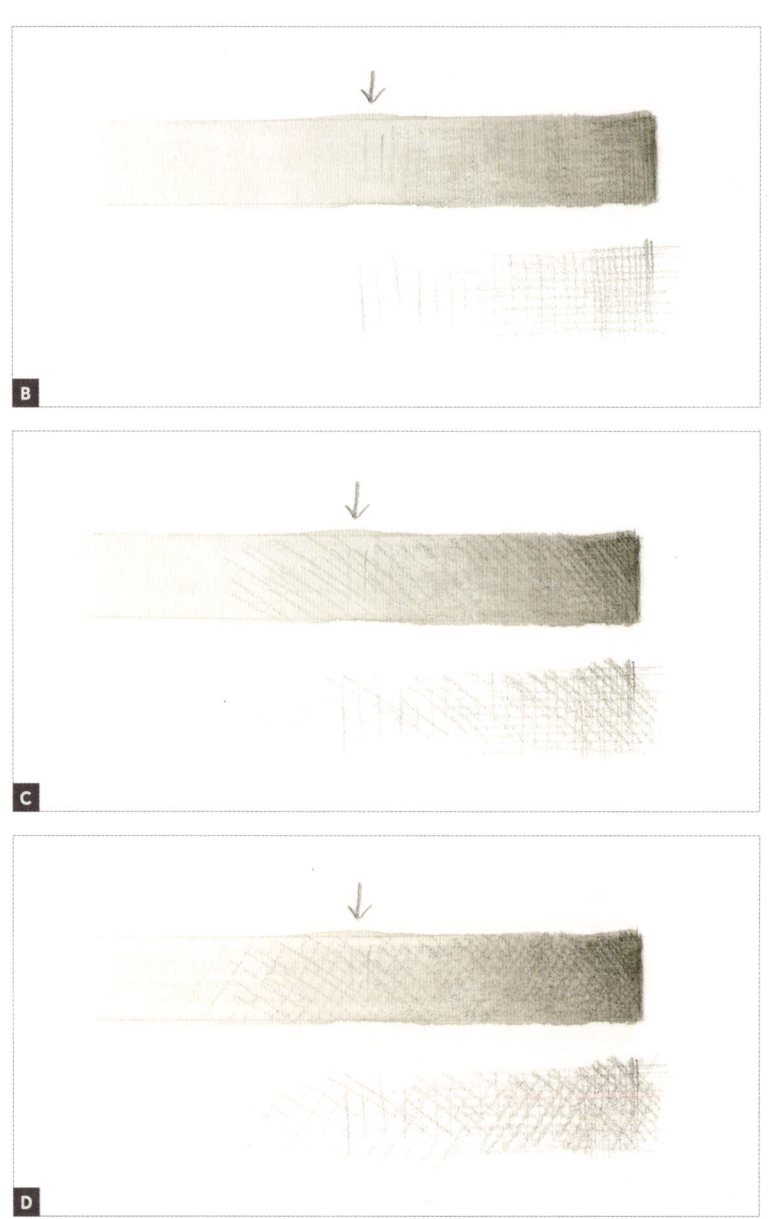

B

C

D

4. Continue to build layers of tone with tiny circular strokes, creating a seamless blend of values from as dark as you can get on the right side to a gradual fade into the lightest area, which is the blank paper color. **(E) (F)**

5. Compare your watercolor tone bar to your nine-step tone bar to make sure you have a complete range of tones. **(G)**

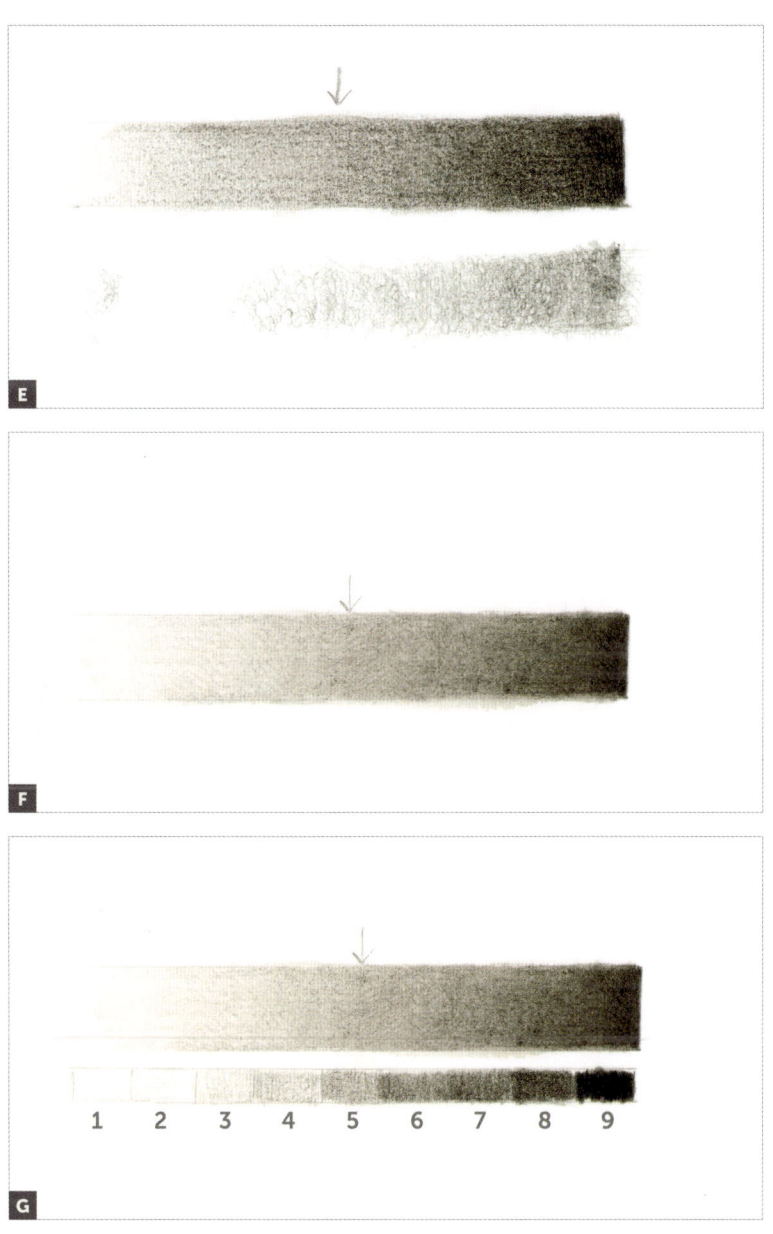

A Curved or Three-Dimensional Tone Bar

ADDITIONAL MATERIALS

- Watercolor brush #2
- Dark Sepia watercolor pencil #175
- Dark Sepia colored pencil #175

WATERCOLOR PENCILS

#175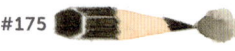

COLORED PENCILS

#175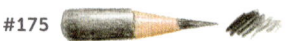

A curved tone bar represents a form that is curved or bending away from or toward light, thus creating various values. Use the same technique for building layers from the previous lesson on a curved tone bar here. This time you'll leave the blank area inside the bar, near the left edge. This lightest area is an imagined highlight. As you build tones to create a bending tone bar, you will start to see the three-dimensional illusion develop right before your eyes. This is a magical feeling, so be sure to go slowly and enjoy the process. I often imagine that my pencil point is a tiny insect crawling across this curved bar, starting in a dark, shadowed valley and slowly creeping up toward the sunlight at the top of a hill as I get to the highlight.

1. Draw a light outline of a curved tone bar with graphite pencil approximately ½ inch high by 2 inches wide. **(A)**

2. Using a #2 brush, apply a graduated layer of graduated Dark Sepia watercolor, from right to left, stopping about ½ inch from the left side. Next apply a small amount of watercolor starting on the left side, leaving a small area of the paper showing. This will become the highlight. Let dry completely. **(B)**

3. Using a sharp Dark Sepia colored pencil and building layers slowly, start to layer value from light to dark. Leave a blank area, about three-fourths of the way across from the right, for the highlight. **(C)**

4. To create a convincing highlight that looks as if it is shimmering light and not just an empty area, continue to add more and more of the lightest values into the empty area so that, by the end, the empty area no longer looks like blank paper but feels like bouncing or shimmering light. Instead of adding marks in a straight line, use zigzag strokes, little dots, or whatever else seems to work. In the end, the amount of blank paper left is small, and there will be several light values bleeding into the highlight area. Observe this enlarged image of a highlight. **(D)**

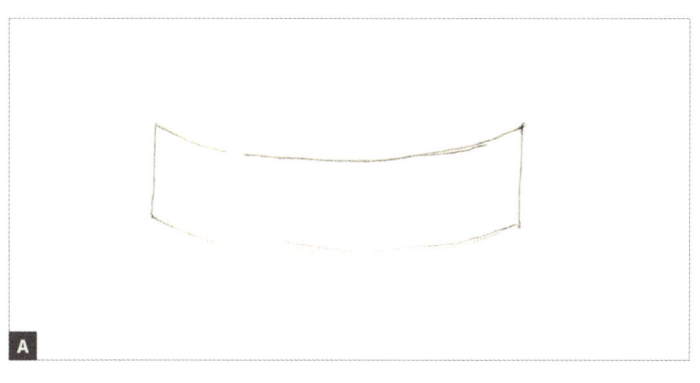

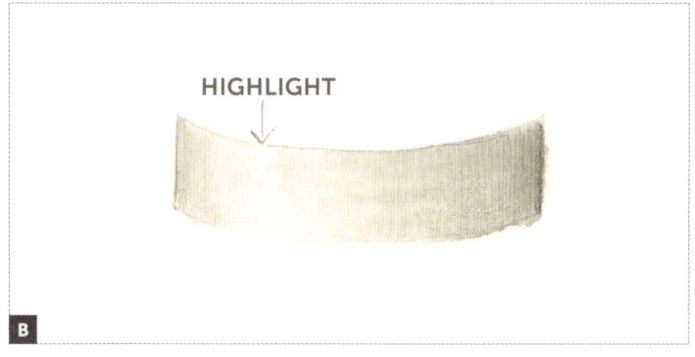

HIGHLIGHT

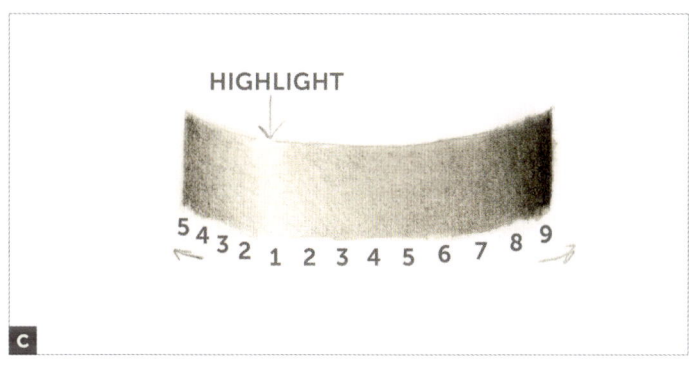

HIGHLIGHT

5 4 3 2 1 2 3 4 5 6 7 8 9

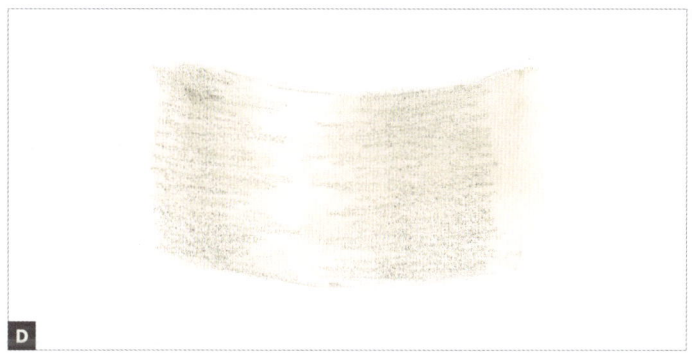

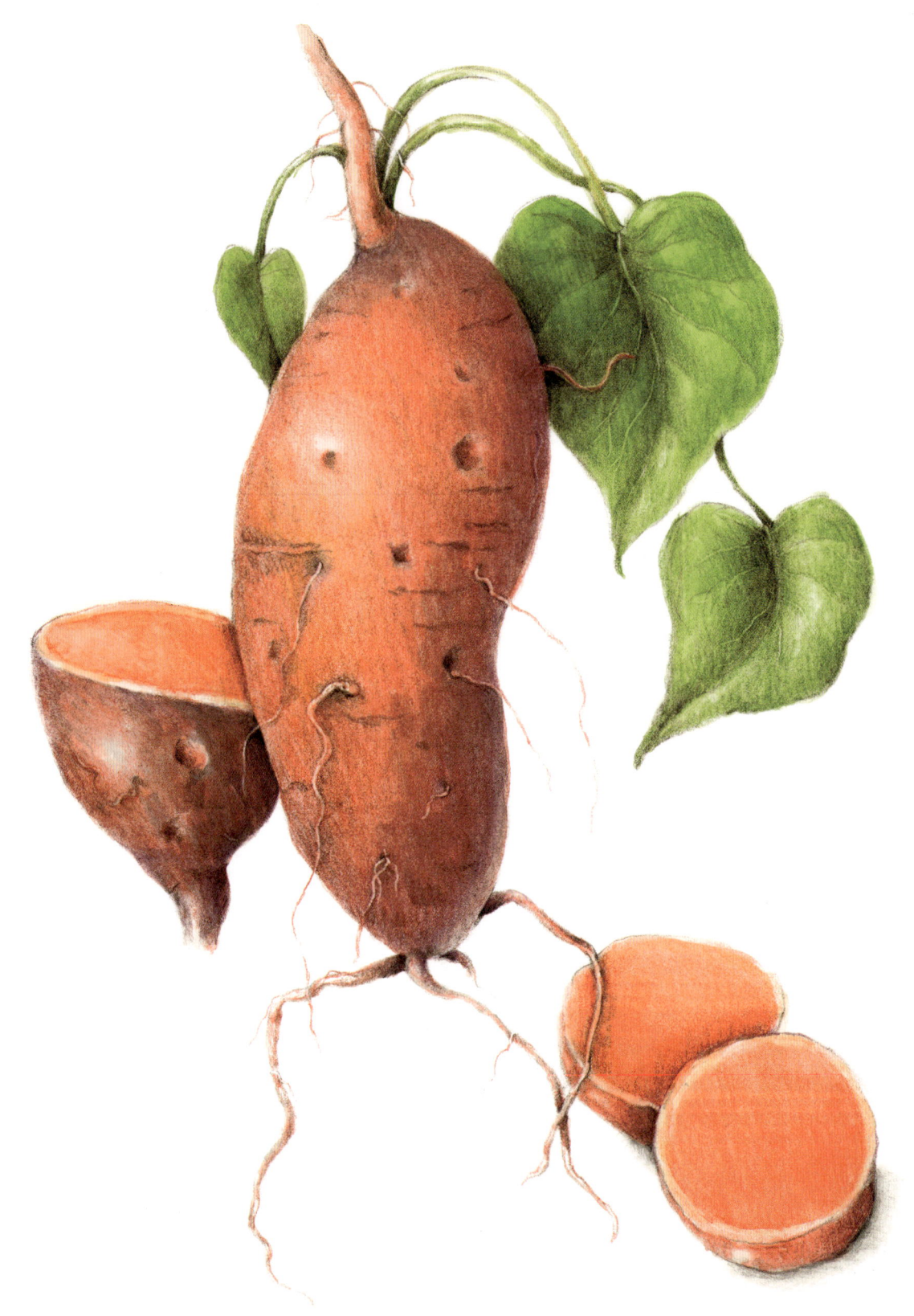

CHAPTER TWO

ADDING TONE TO THREE-DIMENSIONAL FORM

Now that we've practiced the basics of toning, we'll apply this technique on some simple forms in nature. The color of these subjects is fairly neutral, which is helpful. In the beginning, it's best to learn to use light and shadow with neutral tones because it's easier to understand and more important than drawing in color. Cross-contour lines indicate the surface contour, or bending, of a form and help to visualize and describe three-dimensional form more clearly. Cross-contour lines are especially helpful with our imaginary light source as we draw and describe forms bending away from and toward the light. They can also help function as a map when it comes to the placement of light and shadow, or to show complex surface variation, such as on rolling flower petals.

Highlight

A highlight can vary depending on the surface of the subject. If the subject has a shiny surface, the highlight will be much brighter. If the subject has a dull surface, you might not even see the highlight. The important concept to remember is that you're creating the illusion of three dimensions on a flat piece of paper, so you have to exaggerate at times to enhance this illusion. Even if you don't see the highlight, you can draw one in using your head light. The most crucial thing about a highlight is placement. It should be placed to the left or right of center, closest to the light source you are using. Try not to position it in the middle of your form as it will be misleading and not describe your light source correctly. The second important thing is to make sure the highlight doesn't appear to be just a hole or empty space. You want it to shimmer so that it feels like light hitting the form. I was teaching a group of five-year-old students once, describing and drawing a cherry with a highlight. One little boy said, "The highlight is the sun hitting the cherry." He understood exactly!

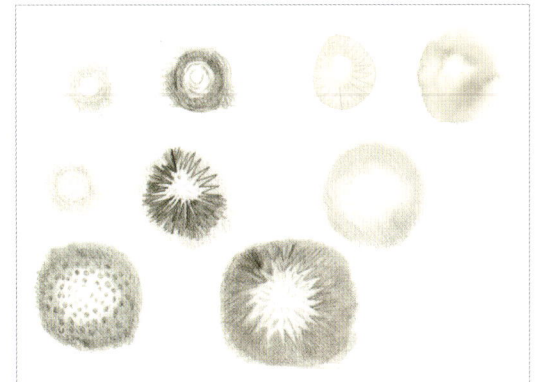

Light some round subjects. Shiny ones are good, because you can better see a highlight; citrus fruits are good, too. Look closely at the highlight with a magnifying glass and analyze the pattern you see. Try to draw the pattern of the highlight, then radiate out into the midtone from there.

Reflective Highlight

Like a highlight, the reflective highlight needs special attention. A reflective highlight is a relatively weak light that bounces off a surface and back onto a subject at the bottom of its shadow side or next to an overlapping area. It's very subtle, and it isn't necessary to always emphasize it. A reflective highlight can be useful to draw when you use a cast shadow, as it shows contrast between the subject and the surface on which it sits. The key to drawing a convincing reflective highlight is to not make it too light. It should be much darker than a highlight: if the highlight is value 1, then the reflective highlight should be about value 5. Also, a reflective highlight is most convincing when it is irregular, diffused, and sort of "glowing," rather than a straight edge. It's good to practice drawing them several times to perfect your technique.

Drawing a Small Tomato in Neutral Tones

Creating round, three-dimensional forms is a very gratifying experience. Once you learn how to make a juicy-looking tomato or a bunch of deep-purple grapes, you may not be able to stop! Drawing fruit is one of my favorite things to do, and the variety of form and color is endless. In this lesson, we'll use a geometric sphere shape as a guide. A sphere is a round form that bends both horizontally and vertically. Cross-contour lines in two directions illustrate this and will help remind you how the form bends away from or toward the light. Look at these curved tone bars vertically and horizontally to help you imagine light and shadow on a round form. Here the toning, or grisaille toning, allows you to do all your thinking about form and light source first and then concentrate on color and details. Next, look at my line drawings that describe the surface contour of a sphere and tomato using cross-contour lines. These lines describe the three-dimensional surface of the form and how it bends away from and toward light in two directions.

SUBJECT

Cherry tomato or other small tomato

ADDITIONAL MATERIALS

- Watercolor brush #2
- Dark Sepia watercolor pencil #175
- Dark Sepia colored pencil #175
- Permanent Green Olive watercolor pencil #167
- Other watercolors to match your subject

WATERCOLOR PENCILS

#175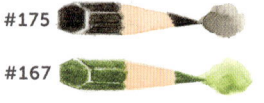

#167

COLORED PENCILS

#175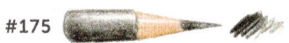

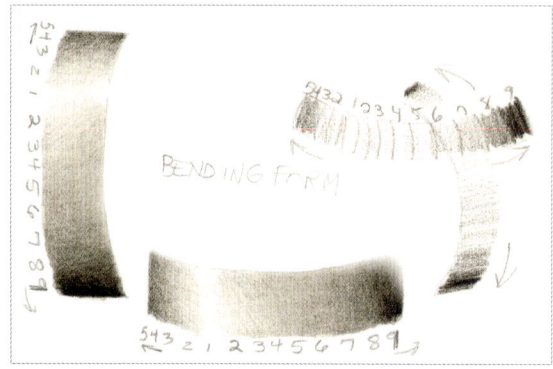

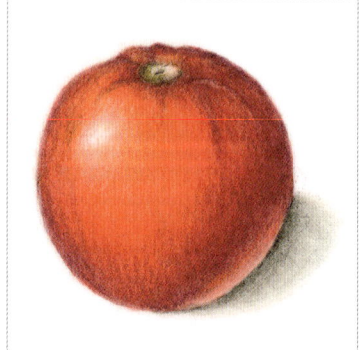

Create your own quick cross-contour line drawings of a round sphere and a small tomato (at left) to help you understand this concept. When choosing how you will draw your tomato, for a view that is easy to complete, choose a view where the stem is hidden on the bottom. For a more challenging view, show the stem side up, and note there is a small, recessed area in the top center of the tomato. I use a medium-size watercolor brush for this lesson because my tomato is tiny. The smaller the brush size, the more control you will have with your watercolor washes.

Repeat this lesson as many times as you want with different round subjects, such as berries, grapes, apples, plums, nuts, and more.

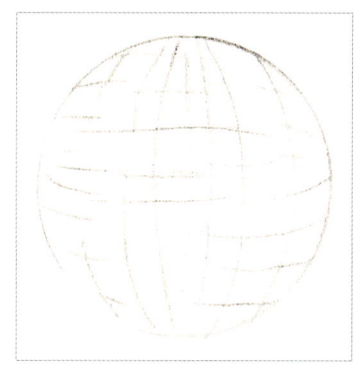

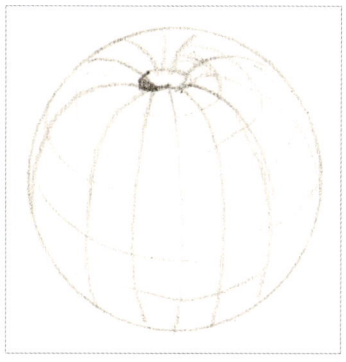

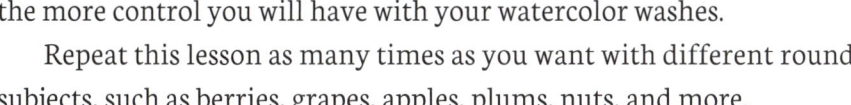

1. Practice visualizing how your head light will hit the sphere and then shine your real light on your tomato as described in the light-source lesson (see page 20). Identify where the highlight and darkest shadows will be placed on your tomato. **(A)**

2. Observe your small tomato and notice how it is similar to a sphere with some slight variations. Choose a view of the tomato to draw. Draw a light outline of the tomato with a graphite pencil. It is fine to lay your tomato on the paper and lightly trace the outline, redrawing after you move your tomato over to make it as accurate as possible. **(B)**

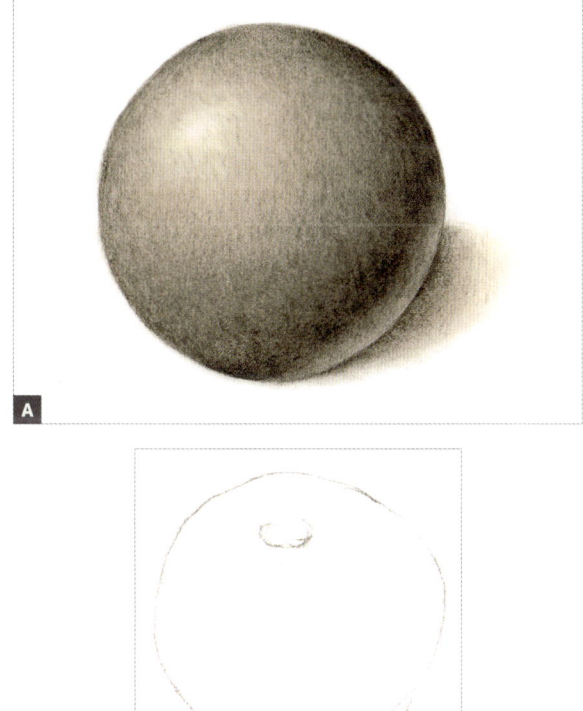

A

B

CONTINUED

3. The recessed area on my tomato is similar to a cone, so I tone it accordingly, creating a shadow in this area. **(C)**

4. Indicate where the highlight and shadow side will be, and, using a #2 brush, apply a light layer of Dark Sepia watercolor to describe the shadow side. Leave the highlight blank. **(D) (E)**

5. Once the watercolor is dry, add a layer of Dark Sepia colored pencil to get a full range of tones from dark to light. **(F) (G)**

6. Try adding a layer of a watercolor wash that is the local color of your tomato. Mine was green, so I applied a wash of Permanent Green Olive watercolor leaving the watercolor off of the highlight. Notice how the underlying shadow tones of Dark Sepia show through, creating a three-dimensional green tomato. In this lesson, the Dark Sepia did all the hard work, then presto! With a layer of the light green wash, the tomato is transformed into a green tomato. **(H)**

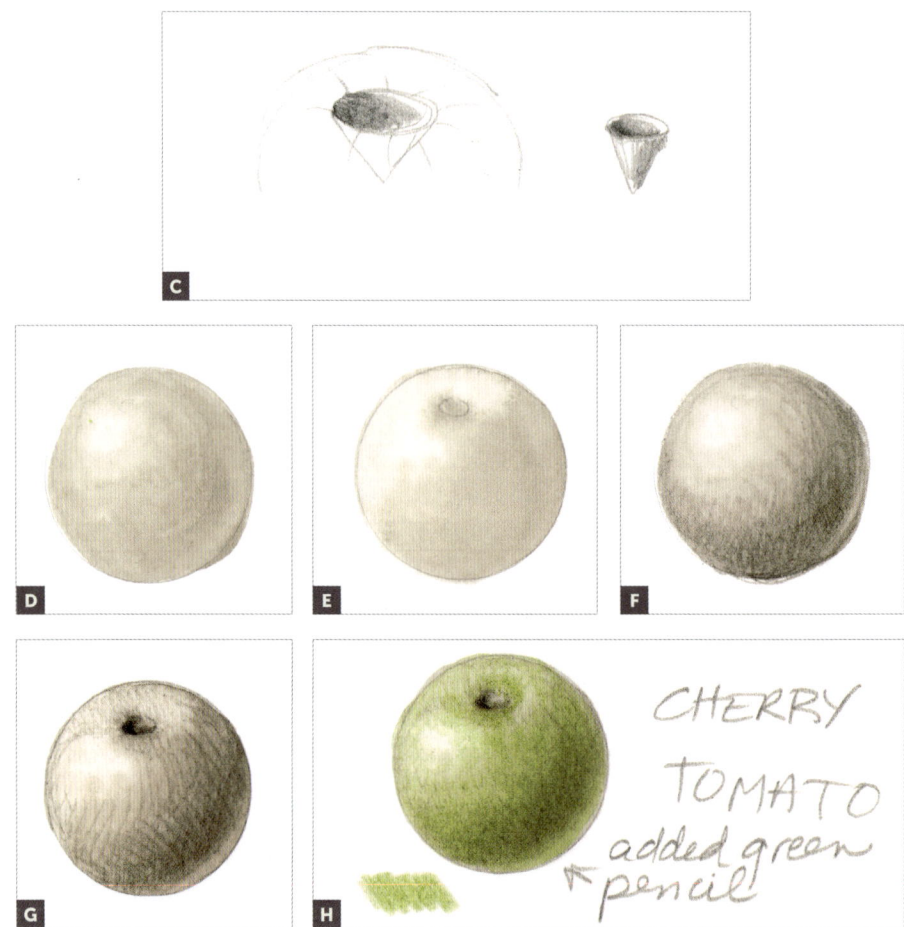

Drawing a Branch

For this lesson, a simple cylinder shape is your guide. A cylinder is a form with straight sides and a circular or bending horizontal cross section. Look at the outline drawing of a cylinder compared with the same cylinder with cross-contour lines to help visualize the surface contour. Straight horizontal lines around the outline of the cylinder shape give the illusion of a flat surface. The curved lines follow the surface contour of the cylinder bending. Finally, look at the cylinder with tonal variation using our ideal light source. Notice how three dimensional and descriptive the cylinder becomes when it is rendered with tones this way.

In this lesson, we'll also tackle burnishing. Burnishing is a way of pressing hard, usually with a light-color pencil, such as Ivory, to blend all the colors together and to remove the texture of the paper showing through on your drawing. After I burnish, I always go back with more of the local color of my subject to intensify the color and also add more shadows if needed. It isn't necessary to burnish over all parts of a drawing. This is a technique that I use when I want a smooth surface and intense saturated color. In dark shadow areas, for example, you might burnish with a dark pencil. Sometimes you might want to leave the texture and strokes of a drawing without burnishing. Repeat this lesson as many times as you want with different subjects that have cylindrical shapes, such as various stems and seedpods.

SUBJECT

Small branch, preferably $1/2$ to 1 inch wide by about 2 to 4 inches long

ADDITIONAL MATERIALS

- Dark Sepia colored pencil #175
- Dark Sepia watercolor pencil #175
- Watercolor brush #2 or #6
- Burnt Sienna colored pencil #283
- Other colored pencils to match your branch's local color
- Ivory colored pencil #103

COLORED PENCILS

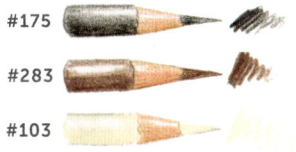

#175
#283
#103

WATERCOLOR PENCILS

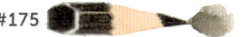

#175

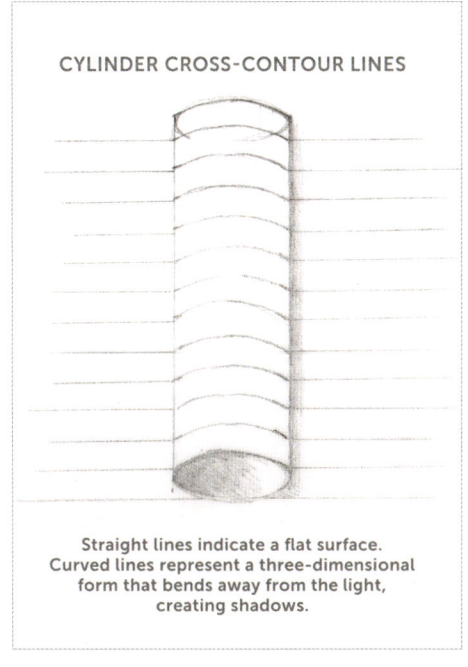

CYLINDER CROSS-CONTOUR LINES

Straight lines indicate a flat surface. Curved lines represent a three-dimensional form that bends away from the light, creating shadows.

CONTINUED

1. Using my drawing as a model, draw a cylinder and add tone to it slowly with Dark Sepia colored and watercolor pencils. This will help you understand the light source model you'll use to draw your branch and will be additional practice on slow toning. The bottom of the cylinder shows the inside hollowed-out area. Add the darkest shading on the right side, and have it get lighter toward the left to indicate this. **(A)**

2. Observe your branch and draw a light, life-size outline of the branch with a graphite pencil. It is fine to lay your branch on the paper and lightly trace the outline, redrawing after you move your branch over to make it as accurate as possible. Position a light on your branch as described in the light-source lesson (see page 20). Look at my drawing of a branch with cross-contour lines to help you visualize its three-dimensional surface. **(B)**

3. To draw the first layer, use a sharp Dark Sepia colored pencil to redraw lightly over the graphite, refining the outline drawing, adding in the details you observe on the branch's surface and perhaps a tiny bit of grisaille toning on the shadow side. A well-drawn subject with a first layer of colored pencil will allow you more control before adding watercolor. **(C)**

4. To create the next layer, first use a #2 or #6 brush to wet the paper inside your drawing of the branch with plain water. Then apply the Dark Sepia watercolor starting on the right side of the highlight and moving toward the highlight. Approach the other side of the highlight and move the pigment toward the highlight, leaving the highlight empty as we did in the lesson on a curved tone bar. **(D)**

5. Once the branch is dry, apply more layers with Dark Sepia colored pencil, and, if the local color of your branch has brown hues, add in some Burnt Sienna colored pencil. Also observe the details on your branch, such as lenticels (which allow the branch to breathe), and draw them in lightly at first. At this stage, you can choose to leave a slightly lighter area on the edge of the shadow side, which will become a subtle reflective highlight.

6. Fill in closer and closer to the highlight. I do this sometimes with small dots, dashes, or irregular zigzaggy toning to make the highlight appear irregular and not an empty stripe. **(E)**

7. Take a look at your subject, and add some additional color where appropriate. Finish your drawing by burnishing with an Ivory colored pencil. Then add more layers of colored pencil to blend, refine, and saturate the branch. **(F)**

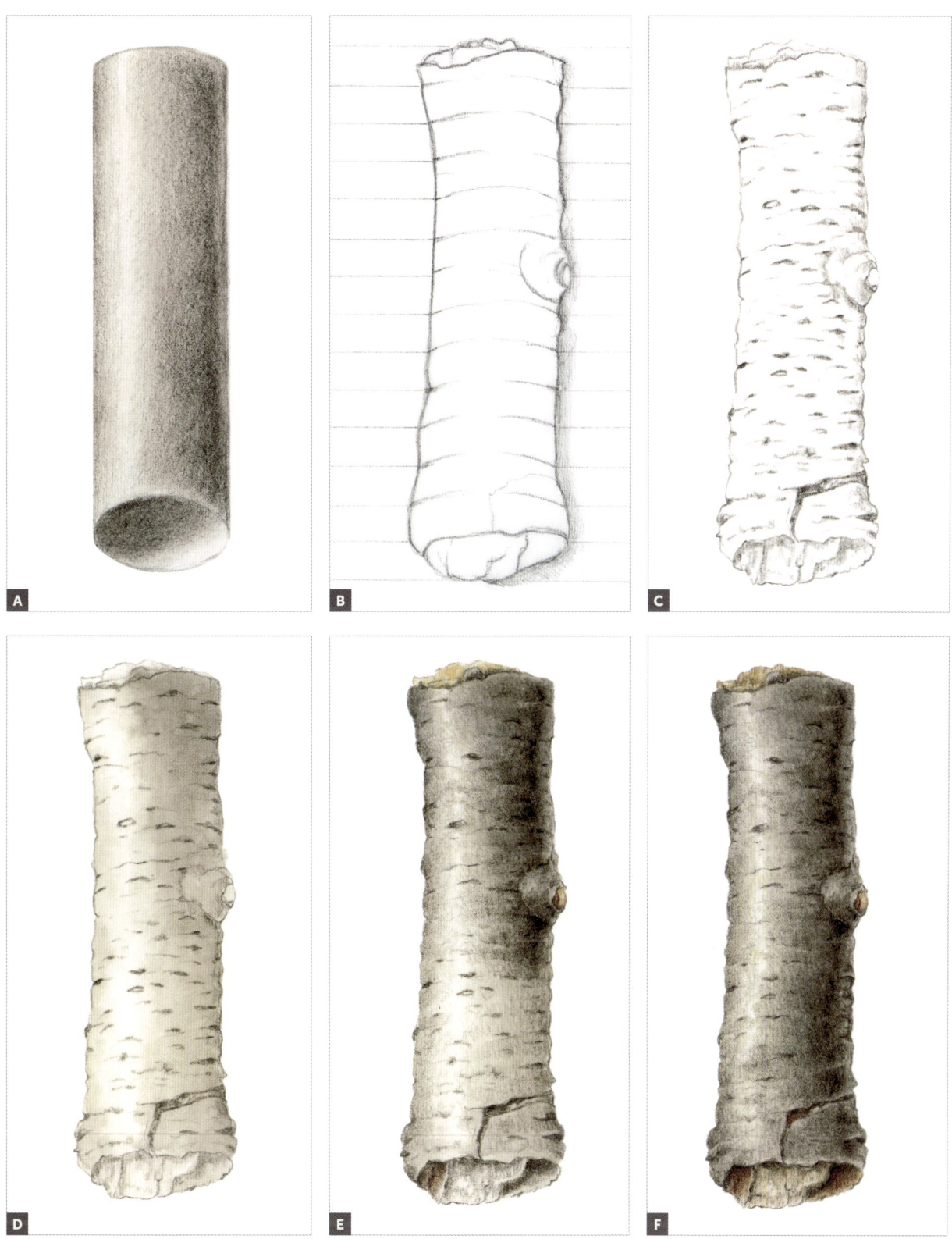

Combining a Cylinder and a Cup Shape to Form a Mushroom

SUBJECT

Mushroom with cap and stem

ADDITIONAL MATERIALS

- Warm Grey IV colored pencil #273
- Tracing paper
- Watercolor brush #2 or #6
- Dark Sepia watercolor pencil #175
- Burnt Sienna watercolor pencil #283
- Dark Sepia colored pencil #175
- Burnt Sienna colored pencil #283
- Other colored pencils to match your mushroom's local color
- Ivory colored pencil #103

COLORED PENCILS

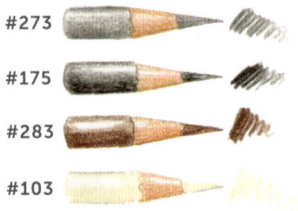

#273

#175

#283

#103

WATERCOLOR PENCILS

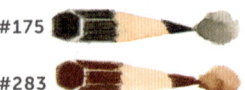

#175

#283

Mushrooms are often a combination of a cylinder and a cup shape. Use these two shapes to help you understand how to draw and shade a mushroom. I love to find and draw wild mushrooms but am very careful not to eat them, as they could be poisonous. If you collect mushrooms in the wild, try not to touch them too much, and be sure to wash your hands after handling. I've started to learn to identify some edible mushrooms with the help of my botanical drawing practice, but I only eat them in the company of highly skilled foragers. The foraged *Amanita muscaria* var. *guessowii* mushroom is poisonous but absolutely fascinating to study and draw. Notice the similarity to the shapes of the store-bought mushrooms used in this lesson, which are much safer but still make a great subject and are readily available. I encourage you to make thumbnail sketches and use them as a reminder of how the light is illuminating your subject, so that you know where to put the shadow side, the midtones, and the highlight.

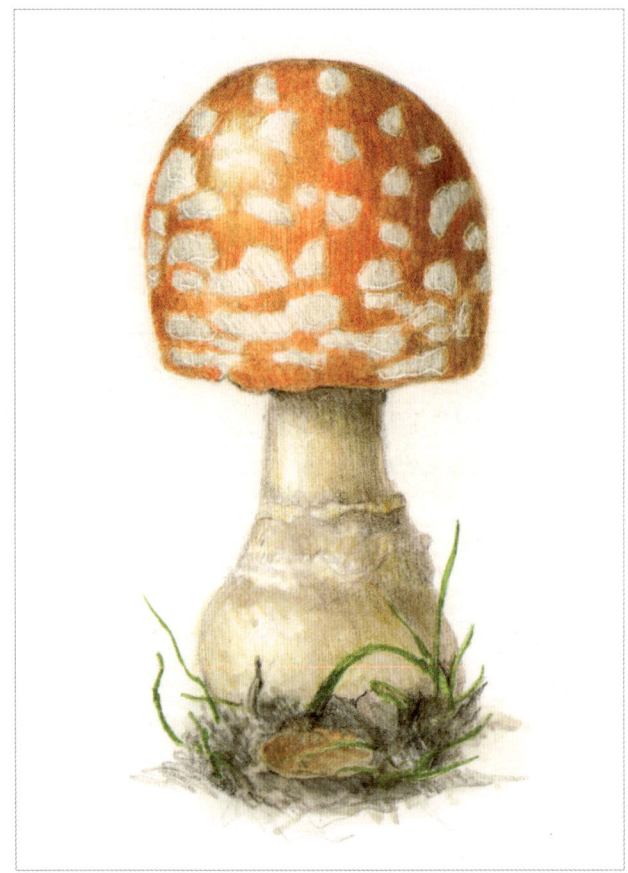

1. Light your mushroom correctly, and note how the light is hitting it. Observe how similar this is to the way the light would hit a cylinder and the top half of a sphere. Pay attention to where your highlight and shadow sides appear. With a graphite pencil, draw a small thumbnail sketch of the lights and shadows falling on your mushroom. **(A)**

2. Measure your mushroom, and draw the outline lightly, life size, with a graphite pencil. If it helps you to lay your mushroom on your drawing to measure, that's fine.

3. Draw over the graphite outline carefully with Warm Grey colored pencil, refining the outline as needed. I always "redraw" my subjects with colored pencil and refine my drawing to make it more accurate and precise. **(B)**

4. Place a small piece of tracing paper over your drawing. Lightly draw cross-contour lines on the tracing paper to describe the mushroom top and the stem to help visualize how the mushroom bends away from and toward the light. **(C)**

5. With Dark Sepia and Burnt Sienna watercolors, mix a light watercolor wash using a #2 or #6 brush. Test your color to make sure it's light. Apply a layer of plain water to wet your paper, painting carefully within your drawing lines.

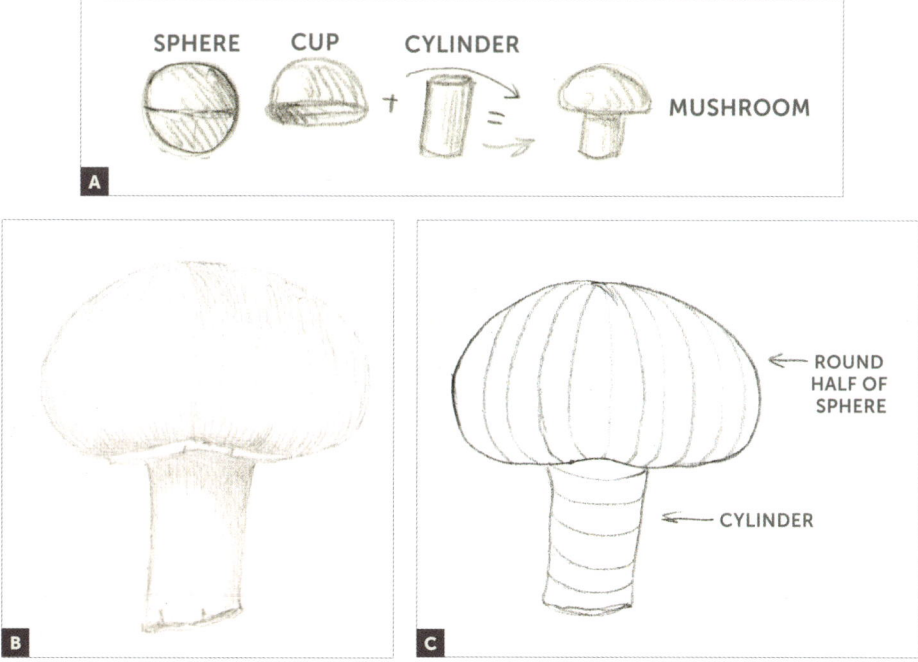

CONTINUED

6. Apply the watercolor wash starting on one side, and move your brush toward the highlight. Then, starting on the other side, paint toward the highlight, making sure you've left an empty spot the color of the paper where the highlight is located. Do this on both the cap and the stem of the mushroom. **(D)**

7. When the watercolor is dry, continue to layer colored pencils to create gradual toning on your mushroom. Use Dark Sepia, Warm Grey, Burnt Sienna, or other neutral colors, depending on the color of your mushroom. Build the layers slowly, and make sure to leave a good highlight. If your mushroom is a white variety, keep the toning lighter and reserve the strong dark tones for the shadow side. Burnish with an Ivory colored pencil and close in on the highlight to make it shimmer; then add any small details. **(E)**

8. For an additional challenge, cut your mushroom in half and add a cross section to your drawing. **(F)**

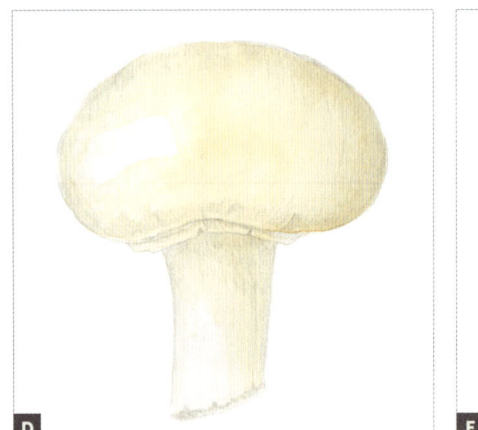

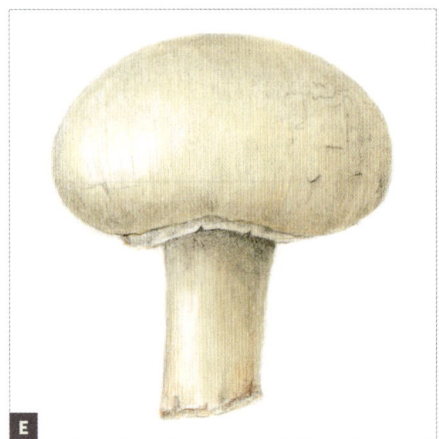

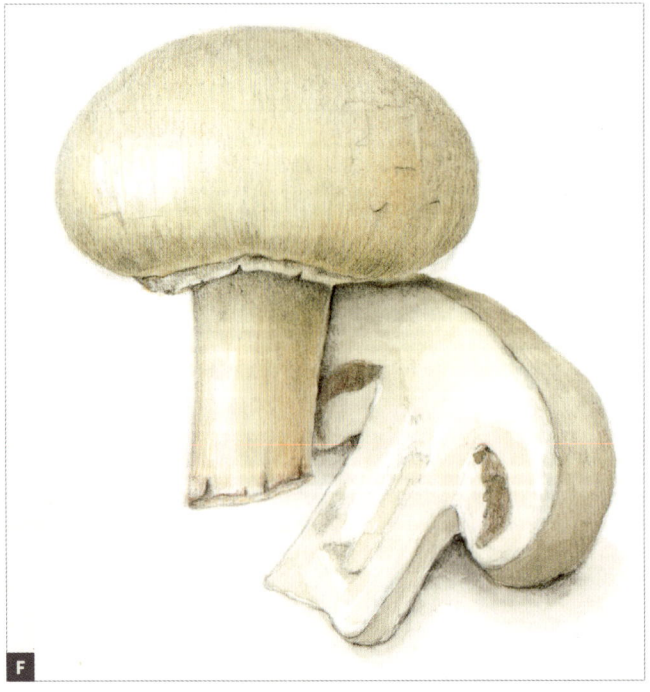

How to Draw Convincing Cast Shadows, Reflective Highlights, and Highlights

A cast shadow refers to a shadow that is created when light is blocked by a subject and results in a dark shadow area on the surface of which the subject is sitting. It may seem obvious what a cast shadow is, but I think cast shadows should be talked about and analyzed much the same way we analyze the subject itself. The cast shadow plays an important role in a drawing: it *describes* the surface on which the subject is sitting. Without a cast shadow, the subject appears to float in space, so drawing one changes this illusion. In two-dimensional still-life art, cast shadows are important because a painting or drawing is describing a three-dimensional space with objects in it.

Sometimes an image describes dramatic light or bright sunlight, so there might be an emphasis on strong cast shadows. In botanical art, the cast shadow is much quieter, if used at all, because the star of the show is the plant depicted and very strong shadows might be overpowering and misleading—most often we see a subject floating on a white background without a sense of space. Sometimes there are elements placed on a surface to emphasize, for example, the difference between a hanging branch of fruit with leaves and stems and a cut fruit sitting on a surface. I like to use cast shadows for this purpose. If you use cast shadows, please be aware of the powerful way they can change or influence your art.

ADDITIONAL MATERIALS

- Gray Verithin pencil
- Warm Grey IV watercolor pencil #273
- Watercolor brush #2
- Black Verithin pencil
- Dark Sepia colored pencil #175
- Ivory colored pencil #103

WATERCOLOR PENCILS

#273

COLORED PENCILS

#175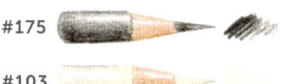

#103

HIGHLIGHT

REFLECTIVE HIGHLIGHT

REFLECTIVE HIGHLIGHT

CAST SHADOWS

CONTINUED

First, realize that a shadow is a shadow and not an object, therefore it must feel like a shadow—not be solid—and disappear into the background. It is the supporting cast, not the star of the show, so it should not steal the spotlight. The main role of the cast shadow is to show that the subject is sitting on a surface, to ground it. Adding cast shadows is fun, but it takes practice to draw a convincing cast shadow that disappears into the background. Follow these steps mostly using Verithin pencils instead of Polychromos, because Verithins have a harder lead and lay down a finer, quieter texture, which is great for keeping shadows soft.

When you draw a round fruit or sphere, make sure to position your light source and note the placement of the cast shadow. With light coming from the upper left, from about a 45-degree angle, the cast shadow will be on the lower right side of the fruit and will create a shadow that goes back in space at about a 45-degree angle as well. The shadow is darkest right where the subject meets the surface on which it sits, but it immediately starts to lighten as it spreads away from the subject and soon disappears seamlessly into the background. Having a reflective highlight on the fruit will give contrast at the edge where the shadow meets the subject, which is important (see page 34).

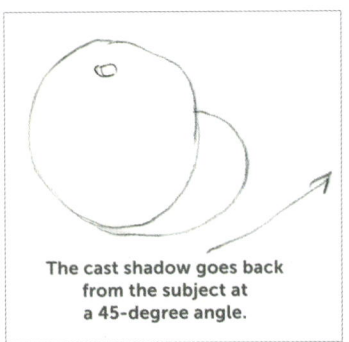

The cast shadow goes back from the subject at a 45-degree angle.

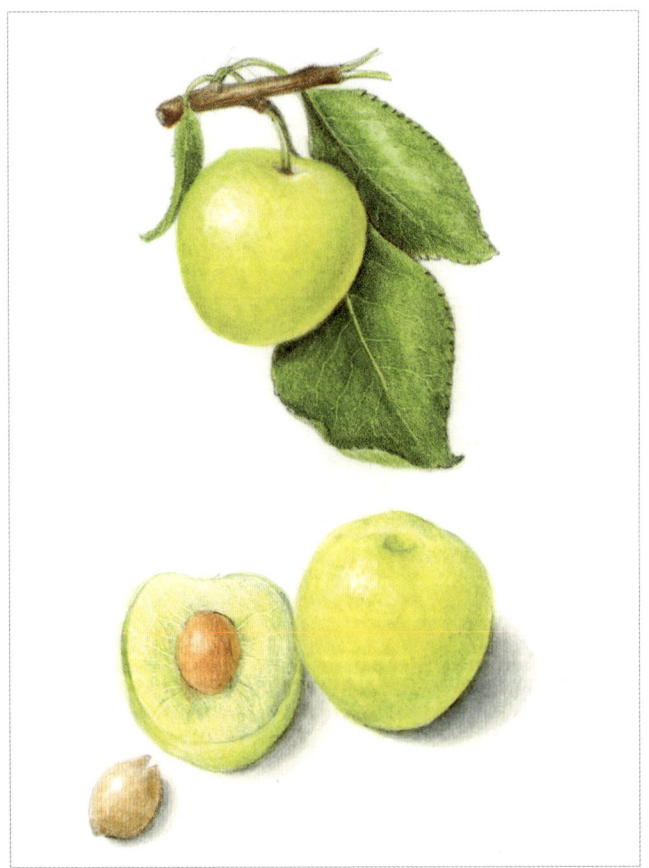

1. Draw a round form, and follow steps from the lesson on drawing a tomato, or use your tomato drawing, to add in a cast shadow. With a Gray Verithin pencil, lightly draw a cast shadow on the lower right that angles back at about a 45-degree angle, and feather it into the background by reducing the pressure on your pencil as you create lighter and lighter marks. Even if your light seems to cast a shadow with a strong outer edge, I recommend that you do not draw it this way but keep it quieter. It should feel like a ghost. **(A)**

2. With a Warm Grey IV watercolor pencil and a #2 brush, create a very light gray wash and apply it over the gray shadow to help soften the shadow and give it a smoother texture. Let it dry before continuing. **(B)**

3. Darken the beginning of the shadow, closest to your subject, with a Black Verithin pencil, and feather the black into the background of your shadow. **(C)**

4. Add a tiny amount of Dark Sepia colored pencil to really darken the beginning of the shadow and create a fading effect.

5. Burnish with an Ivory colored pencil to blend the texture and smooth it out. **(D)**

6. Add tone on the shadow side of your subject, but avoid the outer edge of the fruit or sphere to create a slightly lighter area on the edge. This becomes the reflective highlight at about value 5. Keep it irregular and fuzzy at its edges so that it isn't too strong or straight. Notice how important your reflective highlight will be with the contrast created by the cast shadow. If the reflective highlight seems too bright, tone it down and darken as needed. **(E)**

7. Now look at your highlight. If it feels empty, add more light tones to close in and make it look as if it shimmers. I do this with watercolor first, leaving only one small spot of empty paper and then layering color on top to make the highlight appear irregular. **(F)**

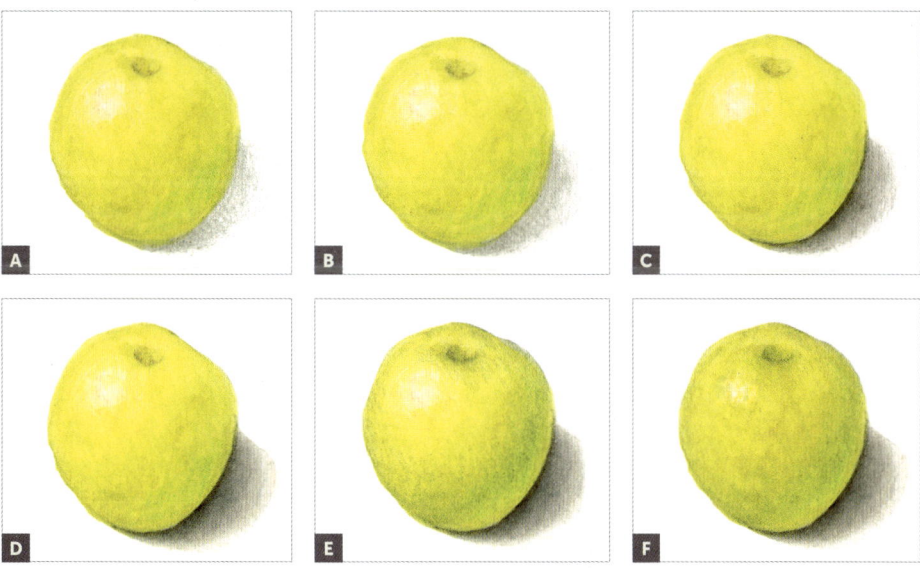

Combining a Cone and a Sphere to Form a Pear

SUBJECT

Brown pear, such as a Bosc

ADDITIONAL MATERIALS

- Tracing paper
- Warm Grey IV colored pencil #273
- Burnt Sienna colored pencil #283
- Burnt Sienna watercolor pencil #283
- Watercolor brush #6
- Dark Sepia colored pencil #175
- Burnt Ochre colored pencil #187
- Other colored pencils to match your pear's local color
- Ivory colored pencil #103
- White colored pencil #101
- Warm Grey IV watercolor pencil #273
- Black Verithin pencil
- Gray Verithin pencil

COLORED PENCILS

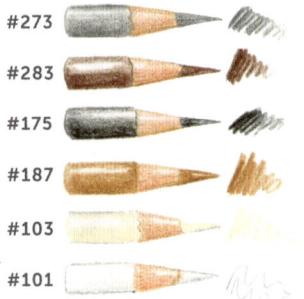

#273
#283
#175
#187
#103
#101

WATERCOLOR PENCILS

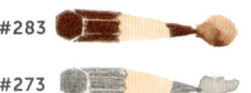

#283
#273

Pears make great subjects, as they are fairly simple in form and color. There are endless varieties, so I encourage you to repeat this lesson with different subjects. Note: In this lesson we are starting with a watercolor wash from the highlight and moving to the edge, the opposite of the previous lesson. I do this so that you experience both ways of working and can choose what feels most comfortable for you. There are many ways to accomplish the same results, and often steps and techniques can be reversed.

1. Light the pear correctly, and note how the light hits it and how similar this is to the way the light would hit a cone and a sphere. Pay attention to where your highlights and shadow sides will be drawn. **(A)**

2. Measure your pear, and draw it lightly, life-size, with a graphite pencil.

3. Place a small piece of tracing paper over your drawing. To help visualize how the pear bends away from and toward the light, lightly draw cross-contour lines to describe the pear shape, and draw a thumbnail sketch for form and light source. **(B)**

4. Add a bit of toning to your pear with Warm Grey IV and Burnt Sienna colored pencils to identify the shadow side and where to leave a highlight blank. **(C)**

5. With Burnt Sienna watercolor, mix a light watercolor wash using a #6 brush. Test to make sure it is light. Apply a layer of plain water to wet your paper, painting carefully within your drawing lines. Apply enough water to wet the pear area well, then let the water soak into the paper for a few seconds.

6. Apply the watercolor wash starting on one side of the highlight and moving toward the outer edge and also around the other side of the highlight, toward the other outer edge and continuing down the pear. The key is to not let any edge dry until you have painted over the whole surface so as not to get any hard edges within the interior of the pear. Do this on the entire pear, leaving highlight on the top (the cone shape) and the bottom (the sphere shape). **(D)**

7. When the watercolor is dry, continue to layer sharp colored pencils to create gradual toning on the pear. Use Dark Sepia, Burnt Ochre, Burnt Sienna, or other neutral colors, depending on the color of your pear.

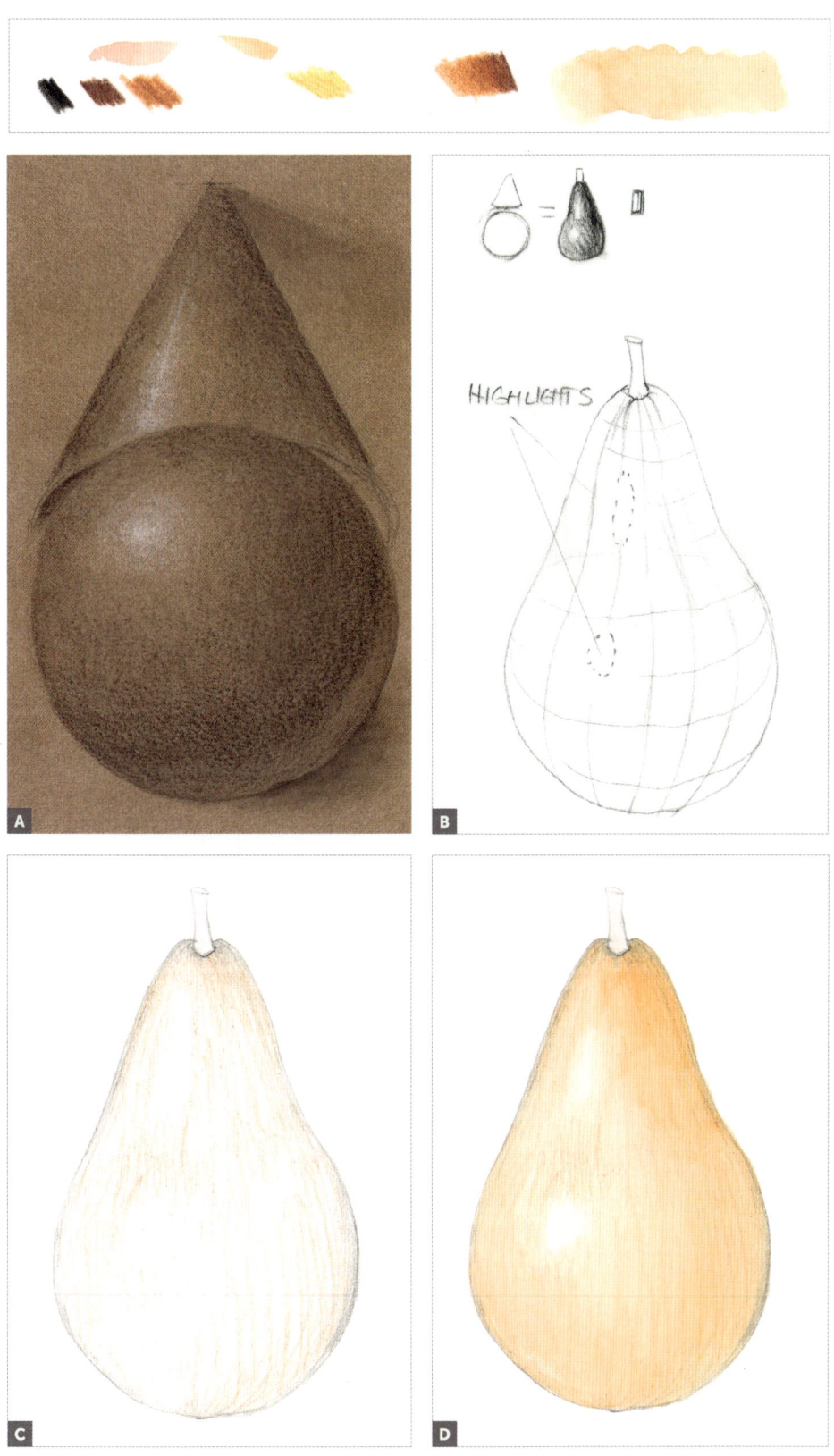

CONTINUED

8.   Build the layers slowly, and make sure to leave good highlights with subtle variations within the highlights. Draw and tone your stem with Dark Sepia colored pencil. **(E)**

9.   Leave a slightly lighter area on the shadow side of the pear for a reflective highlight. Burnish with Ivory and White colored pencils, and close in on the highlight to make it shimmer. Add additional fine details or imperfections. To create a subtle cast shadow, use your brush to apply a light layer of light Warm Grey IV watercolor. **(F)**

10.   Finally, layer Black and Gray Verithin pencils over the shadow to keep it subtle and gradual, and finish with a bit of Dark Sepia colored pencil at the darkest part of the cast shadow. **(G)**

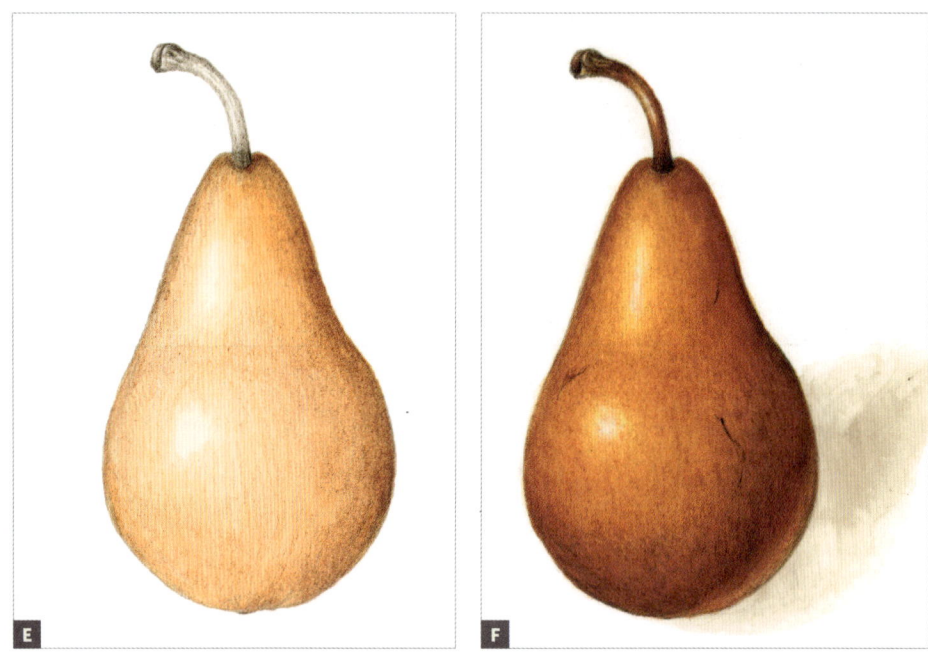

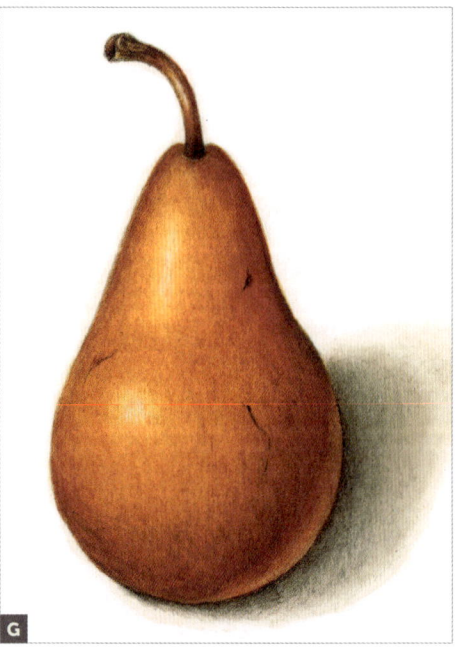

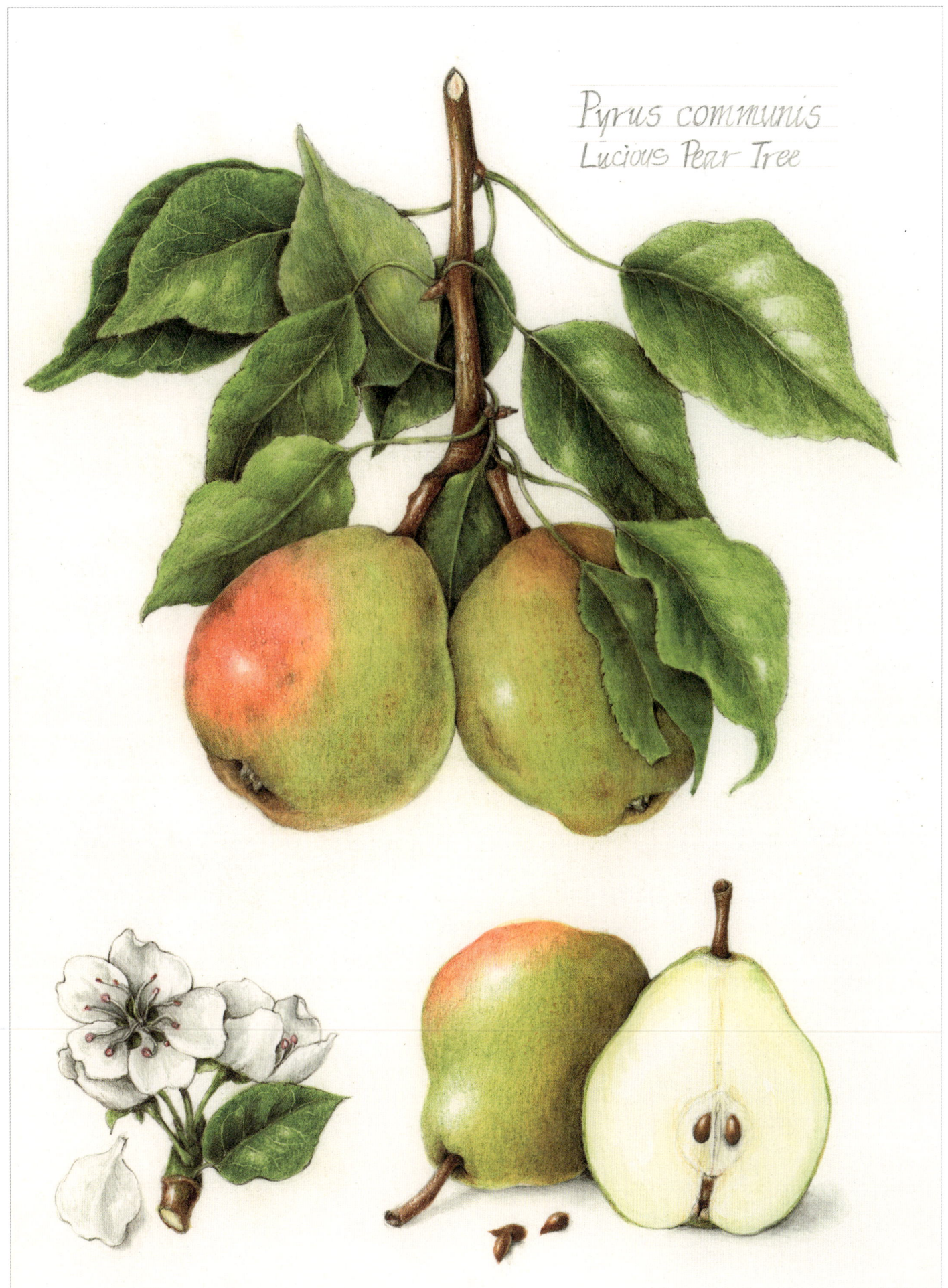

Pyrus communis
Lucious Pear Tree

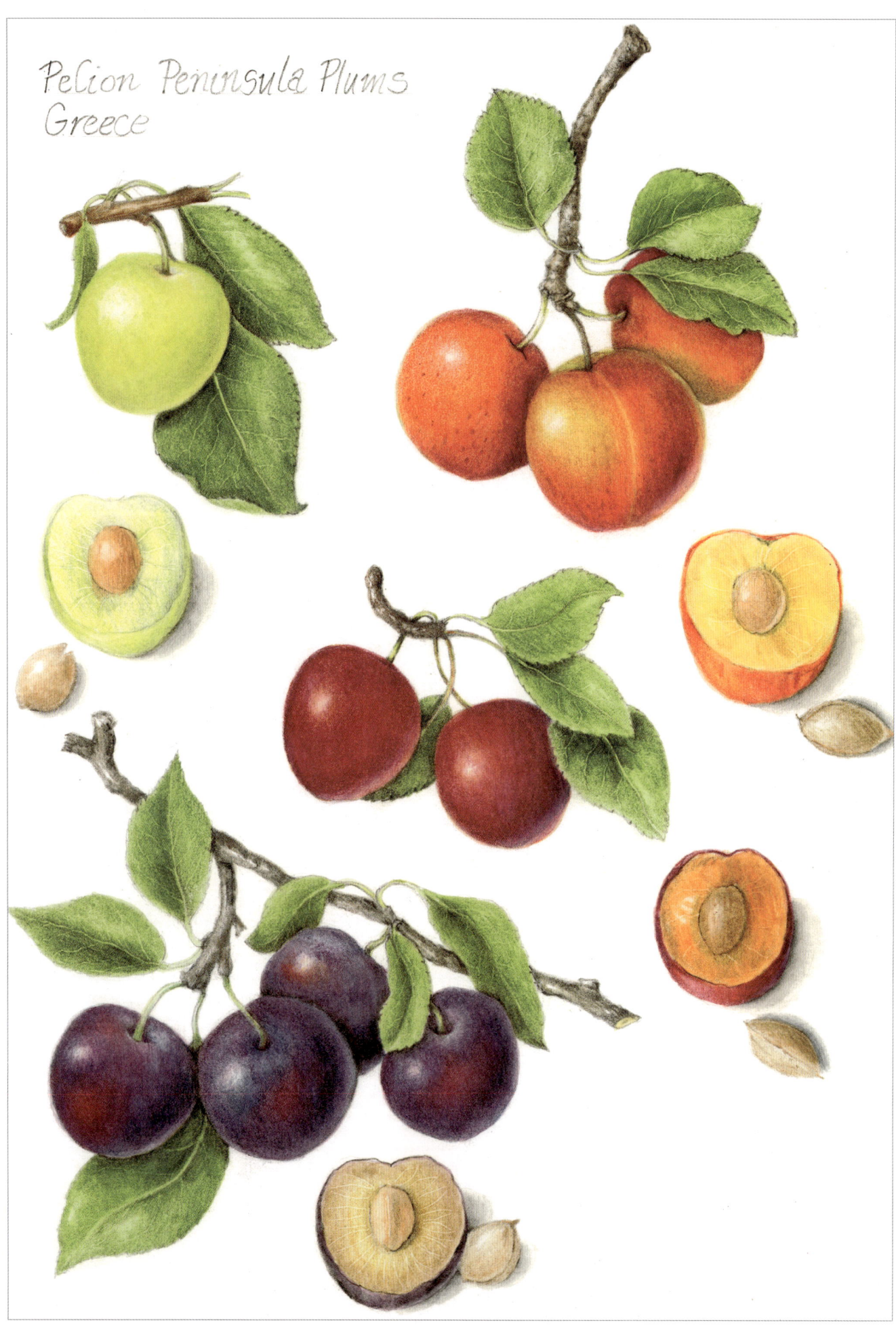

Pelion Peninsula Plums
Greece

CHAPTER THREE

ADDING COLOR TO FORM

The combination of precise color and specific light and shadow variations is crucial to drawing realistic three-dimensional form. We will use nature to learn how to match color, mix variations, and study how light and shadow affect these colors. Then we'll use a color wheel to explore how colors relate to one another, so that by the end of this chapter you'll have mixed a full spectrum of colors as well as their shadows and highlights on a three-dimensional subject. These practical exercises will become your color charts to help you mix colors on just about any form you'll find in nature!

Color Theory: Exploring the Color Wheel

I've always loved looking at color wheels. I liken this to the joy I feel whenever I see a rainbow—nature's way of showing us an arch of colors. A color wheel shows us a complete spectrum of color hues and how each color relates to all the other colors on the wheel. Though we all perceive color slightly differently, a color wheel helps create a universal language that we can all use to understand color and how to mix its subtleties.

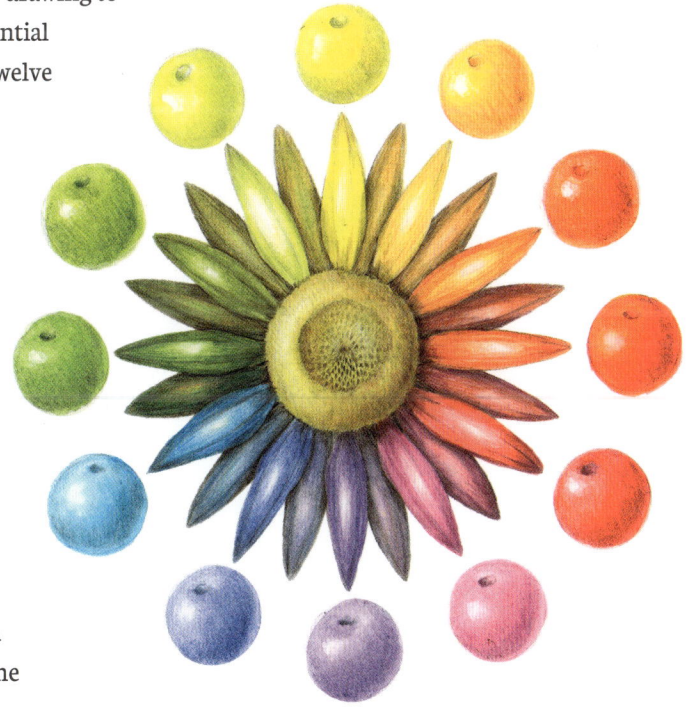

I created the center of my color wheel with a sunflower drawing to illustrate how to use all the pencils that I include in my essential colors. I have organized the top flower petals to show the twelve colors in the color wheel, and additionally, I included petals underneath that are more in shadow, thus are darker and duller in color. For these, I used colors that aren't as bright, referred to as earth tones, that appear often in nature. I've used darker colors for shading in shadow areas and lighter and brighter colors in the highlighted areas of the petals.

For the outside of my color wheel, I was inspired by a real cherry tomato plant, called Toronjina, that was growing in our garden. It is almost the perfect subject for a color lesson. The fruit starts out green, becomes more yellow-green, then yellow, then yellow-orange, then bright orange, all on one stem—a lovely color gradient to practice. Next to my red cherry tomatoes, which change from green to bright red in increments through the

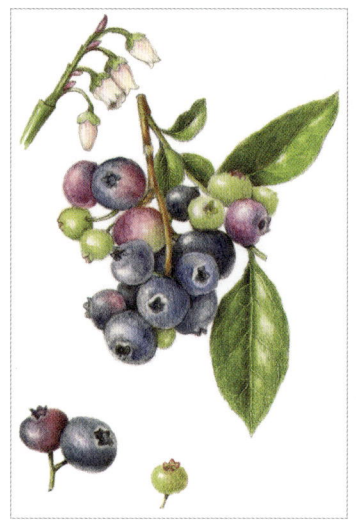

growing season, we have almost a complete range of hues. If only there were tomato varieties in blues and violets, my tomato color wheel would be complete! The drawing of a few blueberries on a bush developing from green to magenta to violet to blue gives us a complete color wheel of round forms.

In describing these various fruits, I use the language of color to explain hue and saturation variations. I'll use a color wheel to illustrate how to mix nature's colors in a realistic way, as well as to create highlights and shadows for each color. This lesson will explain and explore the local color of a form and then create appropriate shadows and highlights for each color.

Let's begin with some basic vocabulary for describing color. Visualizing a color using these definitions can make it easier to mix the color you're trying to achieve.

HUE: the name of a color, also known as local color (for example red, yellow, or blue)

VALUE: how dark or light the color is, as it relates to a nine-step value scale

SATURATION: how bright or dull the color is

My tomato color wheel has twelve distinctly colored tomatoes arranged in a typical color-wheel configuration. I also use double primary color theory, or color bias theory, to create another color wheel, meaning I use two primary colors for each primary color—so I have six primary colors total rather than three. Most of us learned some color theory basics back in elementary school. We were always taught that there are three primary colors—red, yellow, and blue—and that we can mix them together to create all other colors. Mixing two primary colors equally creates a secondary color. These are called orange, violet, and green. The problem with this concept is that pigments used in paints and pencils are never a pure primary color. They will always be biased toward one secondary color or the other. A red can be biased toward orange or toward violet. Having two hues for each primary color allows for precise color mixing so that you can create the color you want, whether it's a bright hue or a dull hue. To create a bright hue, always mix two colors that are biased toward the same secondary color. For a duller color, you can use colors biased toward different secondary colors in different directions.

The final piece of this puzzle is to understand that two primary colors biased in the same direction do not have any of the third primary color when mixed together. When the third primary color is added to a mix, it dulls the color, which is very helpful when mixing colors. Because most of the time we will already be using secondary colors, we won't have to mix them. Keep this biased color theory in mind when you're mixing colors and a color doesn't appear as bright or dull as you want it.

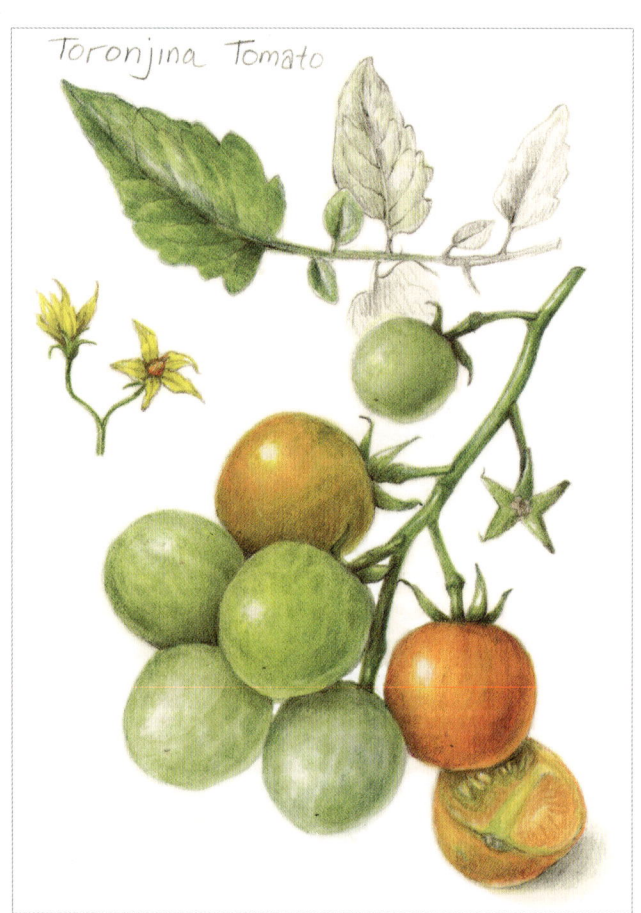

Toronjina Tomato

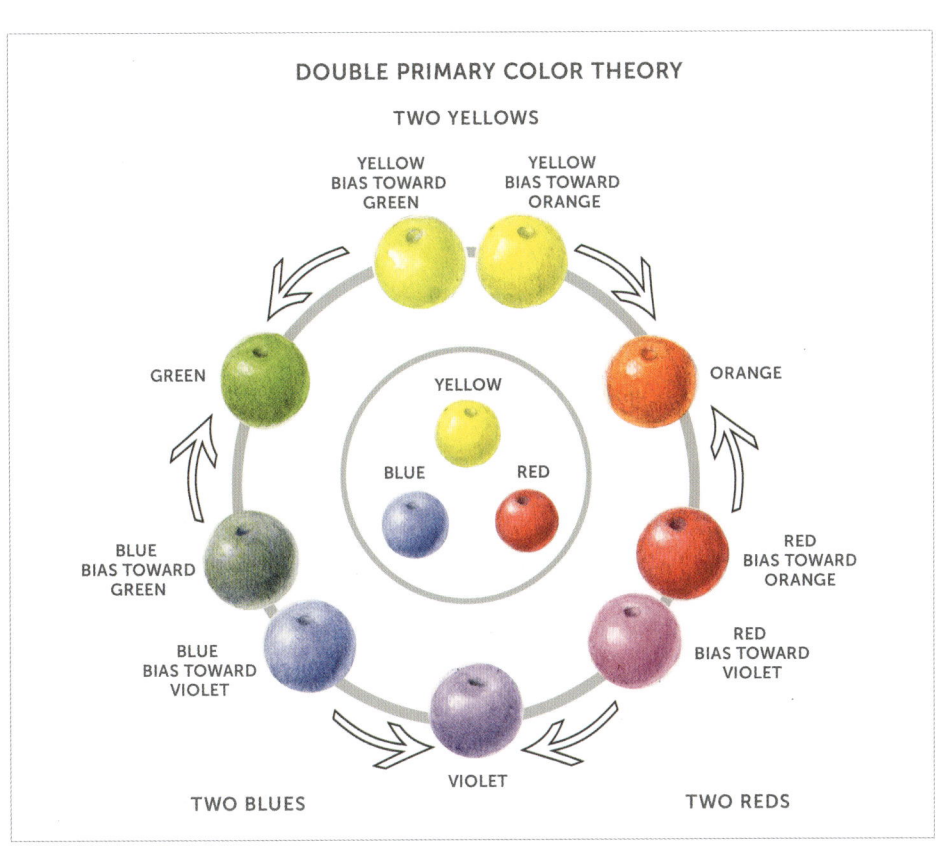

DOUBLE PRIMARY COLOR THEORY

TWO YELLOWS

YELLOW
BIAS TOWARD
GREEN

YELLOW
BIAS TOWARD
ORANGE

GREEN

ORANGE

YELLOW

BLUE   RED

RED
BIAS TOWARD
ORANGE

BLUE
BIAS TOWARD
GREEN

RED
BIAS TOWARD
VIOLET

BLUE
BIAS TOWARD
VIOLET

VIOLET

TWO BLUES   TWO REDS

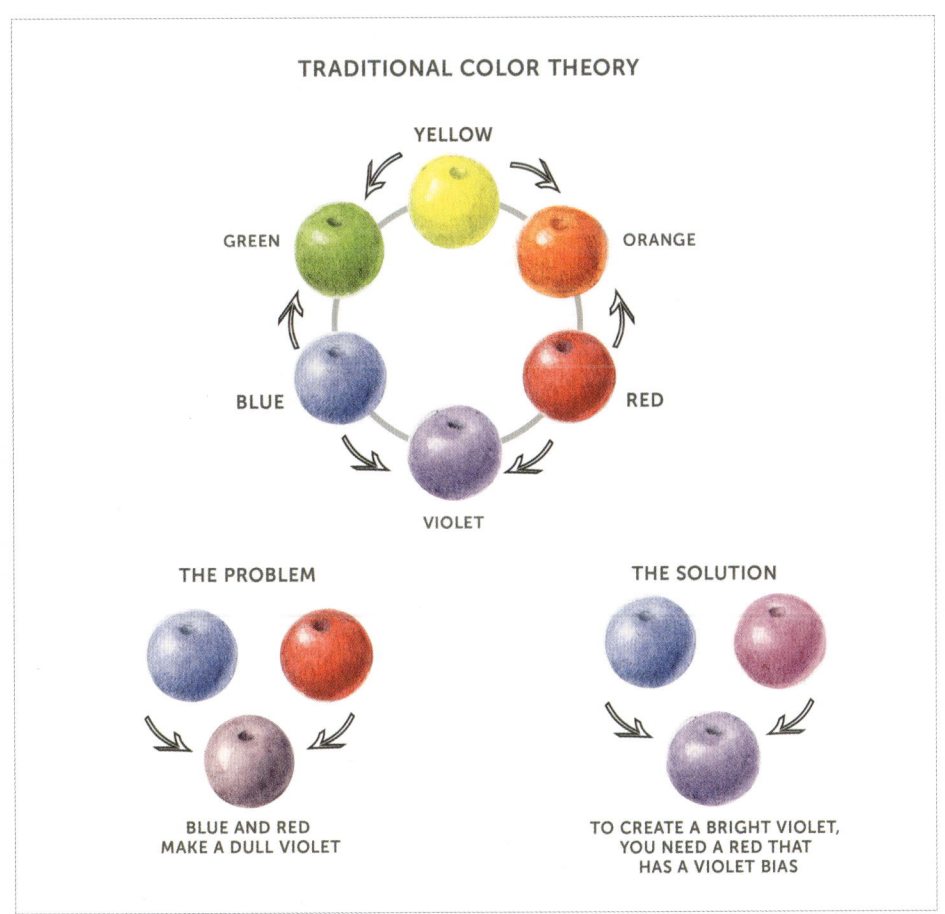

TRADITIONAL COLOR THEORY

YELLOW

GREEN

ORANGE

BLUE

RED

VIOLET

THE PROBLEM

THE SOLUTION

BLUE AND RED
MAKE A DULL VIOLET

TO CREATE A BRIGHT VIOLET,
YOU NEED A RED THAT
HAS A VIOLET BIAS

Primary colors in our set are:

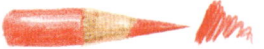

RED BIASED TOWARD ORANGE (Pale Geranium Lake #121)

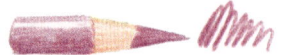

RED BIASED TOWARD VIOLET (Middle Purple Pink #125)

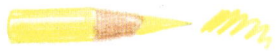

YELLOW BIASED TOWARD ORANGE (Cadmium Yellow #107)

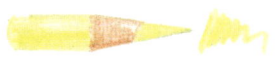

YELLOW BIASED TOWARD GREEN (Cadmium Yellow Lemon #205)

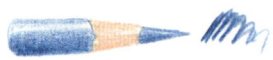

BLUE BIASED TOWARD GREEN (Cobalt Turquoise #153)

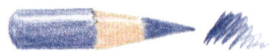

BLUE BIASED TOWARD VIOLET (Ultramarine #120)

Secondary colors are a 50/50 blend of two primary colors. In our set, we have:

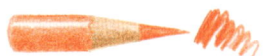

ORANGE (Dark Cadmium Orange #115)

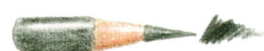

GREEN (Permanent Green Olive #167)

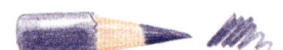

VIOLET (Purple Violet #136)

The other six colors on the wheel are tertiary colors. They are blends of two primaries in a ratio of approximately 3:1. They are called yellow-orange, red-orange, red-violet, blue-violet, blue-green, and yellow-green. This is a visual mixing process and not an exact science. Mix until it looks like the right variation. This is similar to the layering process in creating value steps.

Complementary colors are pairs of colors that fall opposite each other on the color wheel. Any two complementary colors "complete" each other, meaning that together they contain all three primary colors. When you mix two complementary colors together, the result is often a neutral "brownish" color. When you understand how complements mix to create neutrals, you can adjust your drawing to match colors that you find in nature. In the example of a black cherry tomato on the facing page, the pigments in the colors red and green actually mix on the tomato to create a brownish color. Similarly, if you are drawing a green leaf that is beginning to turn brown in autumn, you could add red to the green to neutralize that color and describe the brown you are seeing.

Dark Sepia is such an important shadow color because it's a blend of all three primary colors. Try mixing it with a combination of three primary colors (see the bottom illustration on the facing page).

Usually color wheels show the brightest hues possible, because in color mixing, you can alter a color to make it duller when needed, but it cannot be brightened easily. Starting with the brightest possible hues allows you to dull them when needed. Bright greens are rare in nature, especially on leaves, so I prefer to have my greens less bright on the color wheel, as my color wheel is for practical use. This way, I avoid creating greens that are not realistic in nature.

These series of color wheels can be used as a guide to help you choose and mix colors when you are trying to match the colors in a subject. The next series of lessons in this chapter will walk you through the process of mixing all these color variations.

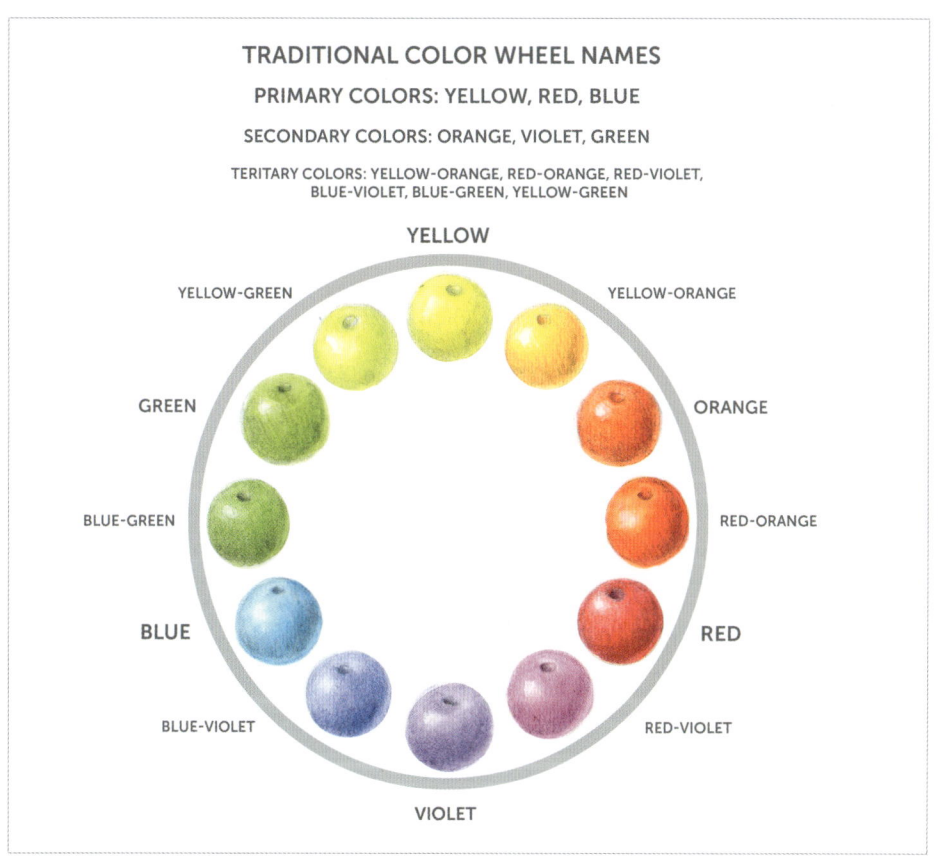

TRADITIONAL COLOR WHEEL NAMES

PRIMARY COLORS: YELLOW, RED, BLUE

SECONDARY COLORS: ORANGE, VIOLET, GREEN

TERITARY COLORS: YELLOW-ORANGE, RED-ORANGE, RED-VIOLET, BLUE-VIOLET, BLUE-GREEN, YELLOW-GREEN

YELLOW

YELLOW-GREEN — YELLOW-ORANGE

GREEN — ORANGE

BLUE-GREEN — RED-ORANGE

BLUE — RED

BLUE-VIOLET — RED-VIOLET

VIOLET

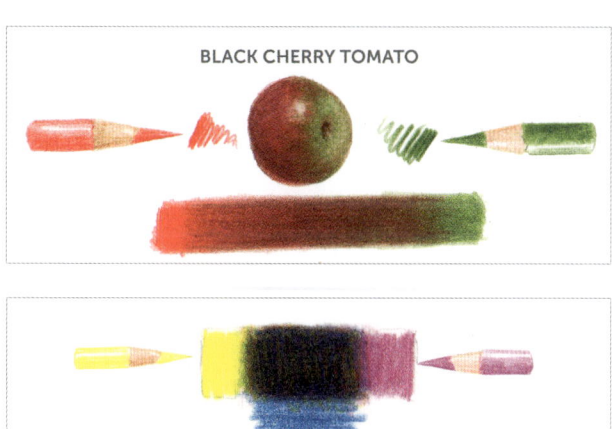

BLACK CHERRY TOMATO

Color Matching and Blending with Colored Pencils and Watercolor Pencils

SUBJECT

Red tomato

ADDITIONAL MATERIALS

- Watercolor brush #2
- Pale Geranium Lake watercolor pencil #121
- Dark Cadmium Orange watercolor pencil #115
- Red Violet colored pencil #194
- Pale Geranium Lake colored pencil #121
- Dark Cadmium Orange colored pencil #115
- Madder colored pencil #142
- Ivory colored pencil #103
- Dark Sepia colored pencil #175
- Cadmium Yellow colored pencil #107 (optional)
- Other colored pencils to match your tomato's local color

WATERCOLOR PENCILS

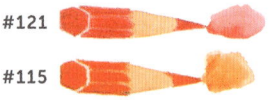

#121

#115

COLORED PENCILS

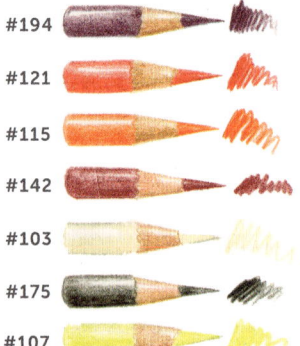

#194

#121

#115

#142

#103

#175

#107

Soon, color matching and combining colored pencils will become second nature. But before attempting to draw a colorful subject, try this helpful practice session of color matching.

The techniques you learn while drawing red objects, such as a shiny red tomato, are applicable to many subjects. Red subjects are challenging because, in addition to matching the main local color of red, you will want your shadow tones and highlight tones to appear realistic as well. The shadow colors need to be dark but still appear red in shadow. Monochromatic tones (only one shade of red from light to dark) are not realistic—convince the viewer that your red tomato is three-dimensional by showing them a range of red hues. This lesson helps create reds that are darker and duller in shadow and brighter and more vibrant in the midtone and lighter areas. Strive for a range of nine values from light to dark.

A cherry tomato is a great first subject because it is easy to find, not too big, and comes in multiples, plus you can eat it when you're finished. Repeat this lesson as many times as you like with subjects of different colors. Once you feel comfortable with this technique and blending colors, it will become easier to pick your colors and do quick, small blend bars to start a drawing with the correct colors. The colors listed here are those that I used, but you should use the colors that match your subject.

1. Place the tomato right next to your pencils to choose colors that you think might combine to create the tomato color. The watercolor pencils should be what you consider the local color of your tomato or what seems to be in the lighter areas of the local color. Using a #2 brush, blend Pale Geranium Lake and Dark Cadmium Orange to create a watercolor wash.

2. Once you've chosen your colored pencils, draw small swatches of each color. Blend your colors together to build an exact color match for your tomato. Then apply a layer of watercolor pencil. Let it dry, then layer colored pencils on top. **(A)**

3. Create two tone bars (see pages 25–31). Make one straight, approximately 2½ inches wide by ½ inch high, and the other slightly curved, approximately the same size, to simulate three-dimensional form.

4. Divide your curved bar and add your first layer of color with your pencils. The 1½ inches on the right will have a complete range of values from dark to light, leaving a bit of the paper showing to indicate a highlighted area. The ½ inch area on the left side goes from midtones to the lightest tone of the paper. Mix up a watercolor pencil base color to lay across the blend bar. Add a bit of water to your brush to get a slightly lighter area of watercolor, and spread it up to the outer edge of the blank highlight of the paper, leaving an empty space as shown. This empty space represents the highlight area. Try to gradually vary your color from light to a midvalue. Do this technique on both tone bars. **(B)**

5. Once the paper is completely dry, add a layer of toning from dark to light using a very sharp Red Violet colored pencil. Draw about four layers of Red Violet, switching the directions of your strokes to create a smooth transition of values. Rather than press hard, build slow layers of tonal variation. Take your time with this and enjoy the process. **(C)**

6. Next, starting on both sides of the highlight area and working out to the darker values, begin adding three to four layers of Pale Geranium Lake colored pencil.

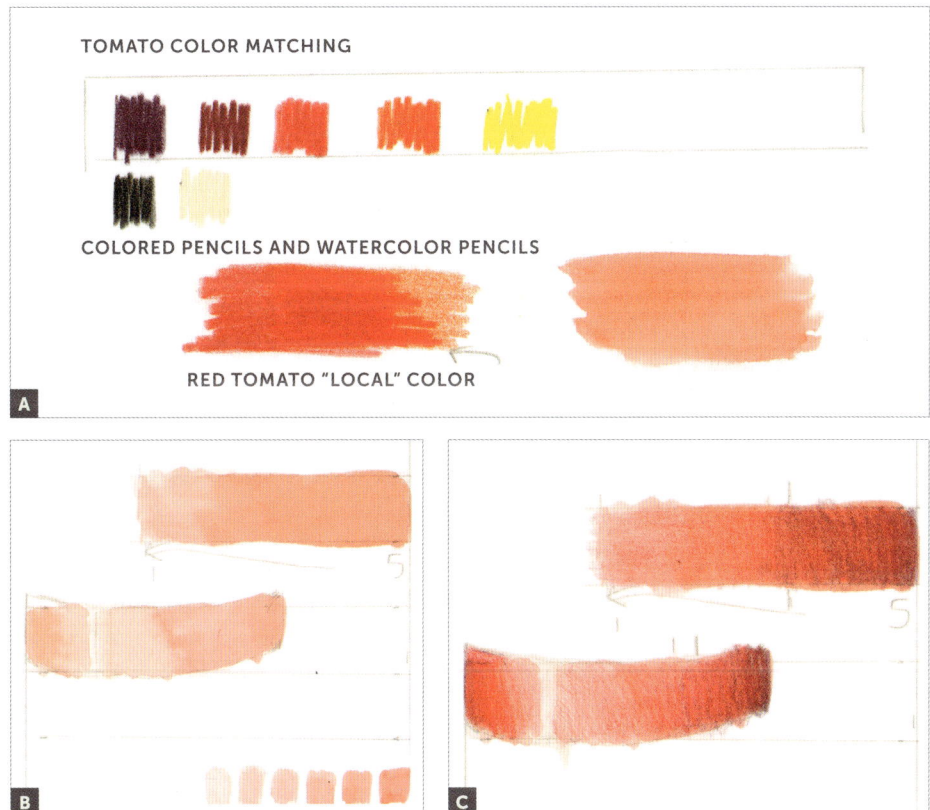

TOMATO COLOR MATCHING

COLORED PENCILS AND WATERCOLOR PENCILS

RED TOMATO "LOCAL" COLOR

A

B

C

CONTINUED

7. Apply three to four layers of Dark Cadmium Orange colored pencil all the way up to the highlighted area, and apply three to four layers of Madder colored pencil, using smaller strokes for smooth blending. Apply the Madder to the midtone areas with gradual toning but not all the way to the lightest areas. **(D)**

8. Burnish the whole subject with a layer of Ivory colored pencil to blend the colors together. Press hard, and try to blend and remove the texture of the paper by blending all the colors together.

9. Reapply one or two layers of all colors on top of the Ivory layer, retaining your color variations. This creates vibrant color while maintaining a smooth transition without any abrupt changes in color and value. To assure that you have dark tones in value 9 at the end of the right side, add a bit of Dark Sepia colored pencil at the darkest side. Consider adding a touch of Cadmium Yellow colored pencil near the highlight. This is optional depending on your tomato. Mine had a bit of yellowish color in it. Place your tomato next to your tone bars and see how well you've mixed and matched the colors. **(E)**

10. If your tomato has a stem, you can practice this blending bar exercise again for the green colors. This time, feel free to make a smaller blended bar. Now move on to the next lesson to draw your tomato using these colors.

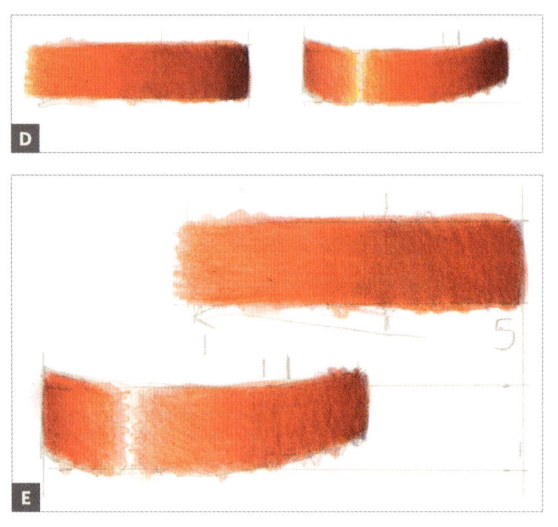

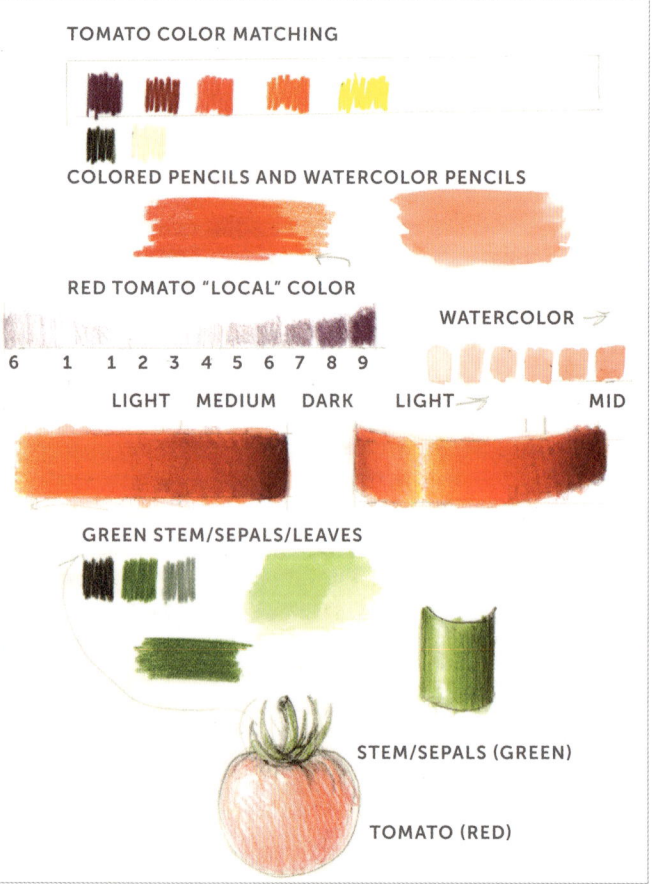

TOMATO COLOR MATCHING

COLORED PENCILS AND WATERCOLOR PENCILS

RED TOMATO "LOCAL" COLOR

WATERCOLOR →

6   1   1 2 3 4 5 6 7 8 9

LIGHT   MEDIUM   DARK

LIGHT →   MID

GREEN STEM/SEPALS/LEAVES

STEM/SEPALS (GREEN)

TOMATO (RED)

Drawing a Red Tomato in Colored Pencil and Watercolor

Now that you've practiced blending and layering your colors, you're ready to apply these techniques to a red tomato. You can use a tomato with or without a stem for this lesson. The goal is to make your tomato look very three dimensional. To achieve this, create a complete range of reddish colors from dark to light. Set up a light source on your tomato so that you can visualize the correct shading. Repeat this lesson again and again with different round subjects, such as small fruits and vegetables and nuts. I've listed the colors I used here, but if you used other colors in the previous lesson and they worked, use those.

1. Measure the height and width of your tomato with a ruler and draw a light, life-size outline of it using an H graphite pencil. It's okay to place your tomato right on your paper to lightly indicate its size and outline, but be sure to always redraw precisely. **(A)**

2. Place a piece of tracing paper over the drawing and draw a cross-contour map to show the three-dimensional surface of the tomato. Indicate the highlight, reflective highlight, and shadow side. **(B)**

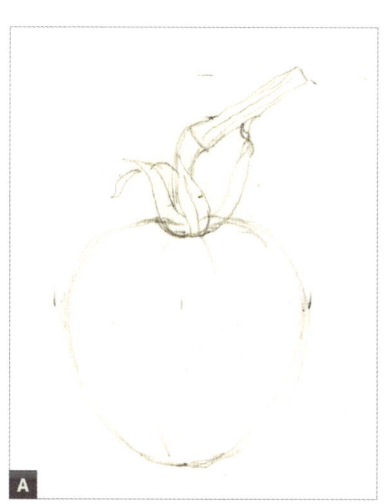

A

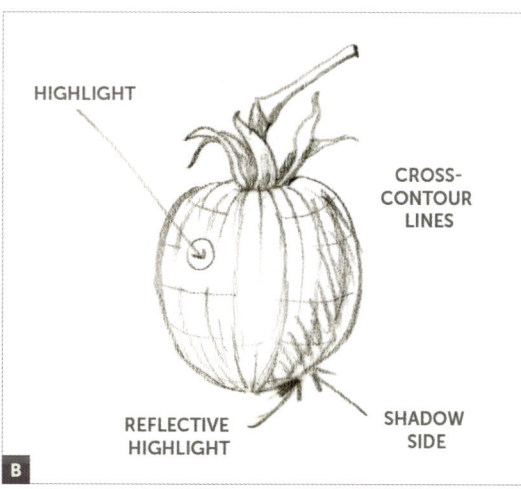

HIGHLIGHT

CROSS-CONTOUR LINES

REFLECTIVE HIGHLIGHT

SHADOW SIDE

B

SUBJECT

Red cherry tomato

ADDITIONAL MATERIALS

- Tracing paper
- Red Violet colored pencil #194
- Dark Sepia colored pencil #175
- Watercolor brush #2
- Dark Cadmium Orange watercolor pencil #115
- Pale Geranium Lake watercolor pencil #121
- Permanent Green Olive watercolor pencil #167
- Pale Geranium Lake colored pencil #121
- Dark Cadmium Orange colored pencil #115
- Madder colored pencil #142
- Ivory colored pencil #103
- Cadmium Yellow colored pencil #107
- Other colored pencils to match your tomato's local color

COLORED PENCILS

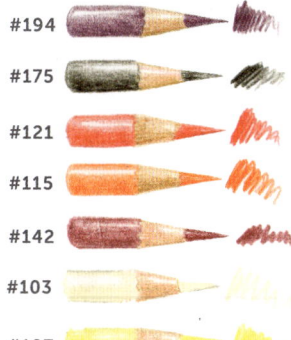

#194
#175
#121
#115
#142
#103
#107

WATERCOLOR PENCILS

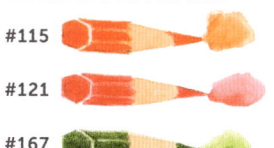

#115
#121

CONTINUED #167

3. Begin the first layer of grisaille toning with Red Violet colored pencil. In previous lessons, I've used Dark Sepia colored pencil for the grisaille layer, but because this is a red tomato, I prefer my grisaille layers to start with a red-tinted neutral shadow color to keep the colors bright. Leave the highlight blank, and make sure to create a seamless blend of dark to light values. Don't press hard at this stage. Consider applying tone with tiny circular strokes for smooth texture application. Remember to always use very sharp pencils to create seamless blends of color and value. **(C)**

4. If your tomato has a stem, apply a grisaille layer with Dark Sepia colored pencil on that area, and indicate any small shadows you see. (See page 43 for more about creating these overlapping shadows.)

5. With a #2 brush, paint a layer of watercolor to create a red-orange color base, leaving the highlight empty. (I used a mixture of Dark Cadmium Orange and Pale Geranium Lake for my watercolor layer.) Always try to match a light to midtone area of the local color for your watercolor layer. When you approach the highlight, dip your brush in plain water, blot on a paper towel, and smooth the transition of the highlight so it is not a hard edge.

6. Apply a watercolor layer using Permanent Green Olive on the stem and any sepals (the small green leaves around the stem). Let this watercolor dry completely. **(D)**

7. Next, start layering Pale Geranium Lake colored pencil, Dark Cadmium Orange colored pencil, and Madder colored pencil (applied lightly) on top, and continue to develop a good shadow side, good midtones, and a blank highlight. Strive for a complete range of seamless values from dark to light. **(E)**

8. Continue to layer these three colors, and start to press a bit harder in the shadow side of the tomato. **(F)**

9. Lightly apply a layer of Red Violet colored pencil in the shadow area, leaving a reflective highlight on the edge of the tomato if desired. **(G)**

10. Add a layer of Ivory and Cadmium Yellow colored pencils in the light-value areas. Do not apply yellow in the shadow areas. Burnish and blend the colors together to remove the texture of the paper and smooth transitions. Apply Ivory colored pencil close to the highlight, leaving tiny areas of the paper without any pencil, but breaking up the area a bit with Ivory to create a shimmery highlight.

11. Reapply all colors to brighten and retain good shadows and contrast. **(H)**

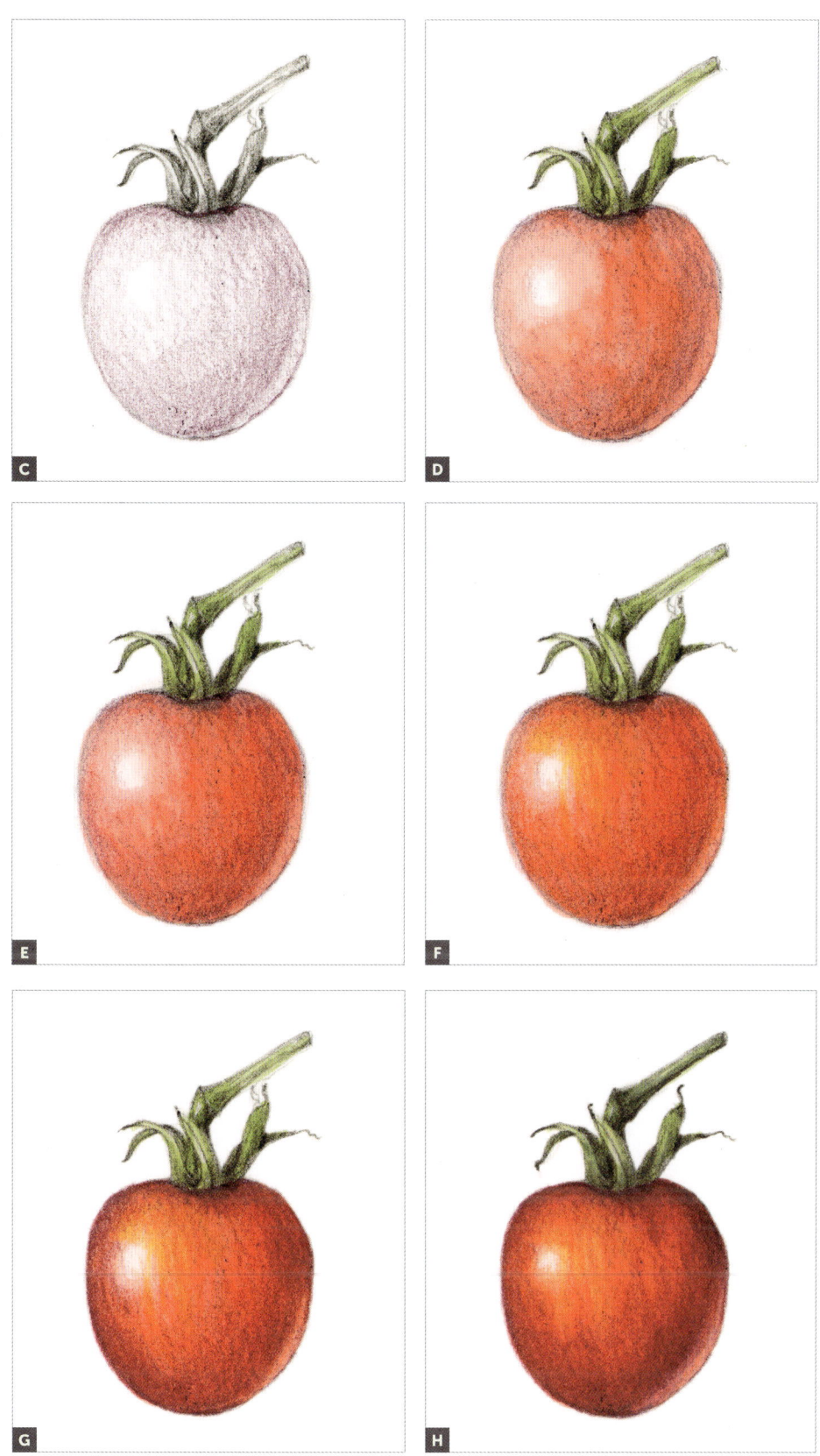

C

D

E

F

G

H

CONTINUED

DRAWING A CROSS SECTION OF FRUIT

For a challenge, cut your tomato in half and add a cross section to your composition. Follow all the steps again for the rendering techniques, and refer to page 86 on measuring in perspective using ellipses. Notice how subtle the cast shadow is on the cut-open fruit. I drew my cast shadow with a light layer of Warm Grey IV watercolor, and I depicted the outer edge of the shadow gradually fading into the background. **(A)** When the watercolor was dry, I layered Verithin pencil Cool Grey 70% and Verithin Black right where the shadow touched the leaf edge. I finished with just the slightest bit of Dark Sepia colored pencil in the darkest area of the shadow. **(B)**

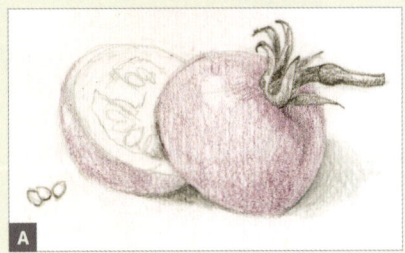

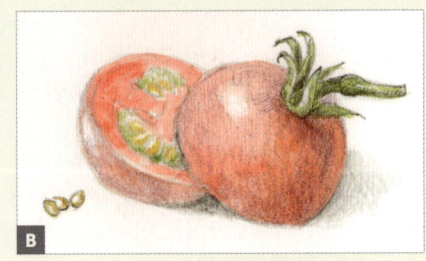

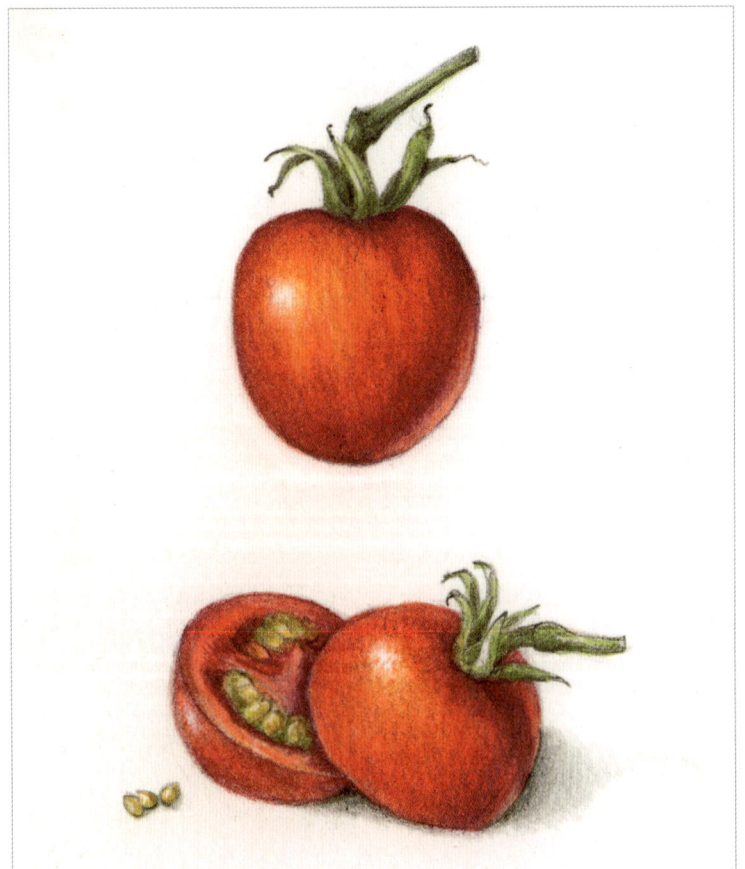

Mixing Bright Colors and Shadows

This lesson combines the techniques you've learned for rendering a three-dimensional form with the basics of color theory. You'll practice combining all the colors on a color wheel with appropriate lights and shadows, and this can be a guide to help you find appropriate colors for any subject in the future. I have not included Dark Sepia in the suggested colors for each sphere because I want you to experience color first. Sometimes too much Dark Sepia creates overly dull colors, so use it cautiously. You can add a touch of it on your shadow sides if you need a slightly darker shadow, especially on colors that have a darker value in their local color, such as blues and greens.

Note: This is a long lesson, and there is a lot of information to absorb. I recommend doing it over several sessions. If you complete your watercolor layers for all of the spheres first, they'll be nice and dry by the time you're ready to add your colored-pencil layers. In this lesson, the step-by-step procedure for creating each round form (steps 1–4, below) is the same, but you will substitute the appropriate colors from one form to the next. This is the same technique used in Drawing a Red Tomato on page 59. Take your time, and pay attention to the way colors blend with each other and how quickly you can make color variations as needed.

ADDITIONAL MATERIALS

- Watercolor brush #2
- All colored pencils and watercolor pencils

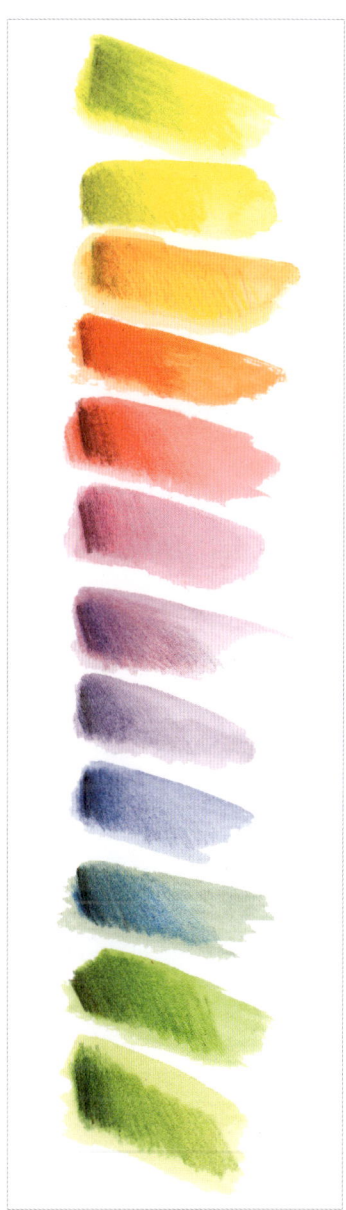

1. With graphite pencil, lightly draw a small tomato or sphere. You can have a stem indentation if you like.

2. Wet the tomato with water, being careful to stay within your outline. While the sphere is still wet, mix a watercolor wash on your palette and paint a layer to create a round form. Leave a highlight of blank paper in the appropriate place. Let the watercolor dry completely.

3. Once dry, add several layers of colored pencil to the round form. Choose appropriate shadow colors, and add shadows to create a three-dimensional form.

4. Saturate your form with colored pencils of the local color, more shadow colors if needed, and a touch of Dark Sepia lightly blended into the shadow, especially when working with the darker colors (such as blues and greens).

CONTINUED

YELLOW, BIAS TOWARD ORANGE

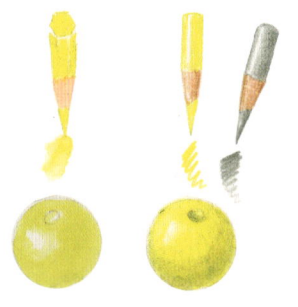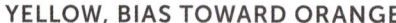

WATERCOLOR PENCIL: Cadmium Yellow #107

COLORED PENCIL SHADOW COLOR: Earth Green #172

LOCAL COLORED PENCIL: Cadmium Yellow #107

Use a light shadow color because yellow is a very light hue. Earth Green is a dull green and works well for this purpose; it's composed of yellow, blue, and a bit of red. You can try mixing this color to understand its components.

YELLOW-ORANGE

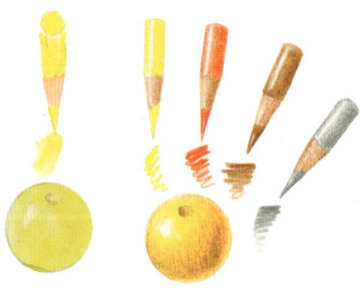

WATERCOLOR PENCILS: Cadmium Yellow #107 and Dark Cadmium Orange #115 (3:1)

COLORED PENCIL SHADOW COLORS: Burnt Ochre #187 and Earth Green #172

LOCAL COLORED PENCILS: Dark Cadmium Orange #115 and Cadmium Yellow #107

Orange has a darker color value than yellow; avoid getting an overpowering orange by adding less orange and more yellow.

ORANGE

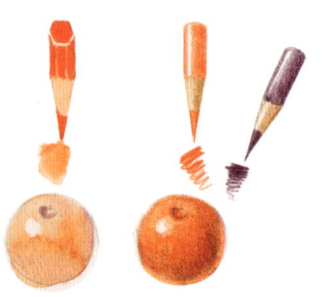

WATERCOLOR PENCIL: Dark Cadmium Orange #115

COLORED PENCIL SHADOW COLOR: Red Violet #194

LOCAL COLORED PENCIL: Dark Cadmium Orange #115

For additional practice, mix Cadmium Yellow #107 and Pale Geranium Lake #121 (1:1) to create an orange color to use as a base coat.

RED, BIAS TOWARD ORANGE

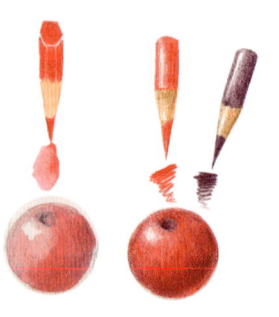

WATERCOLOR PENCIL: Pale Geranium Lake #121

COLORED PENCIL SHADOW COLOR: Red Violet #194

LOCAL COLORED PENCIL: Pale Geranium Lake #121

Observe a real red tomato to decide on the shadow color. I chose a deep, dull wine color (Red Violet) for my shadow. It has red in it with some yellow and a touch of blue. Try mixing a color that matches Red Violet if you want to understand the components of this color.

RED, BIAS TOWARD VIOLET

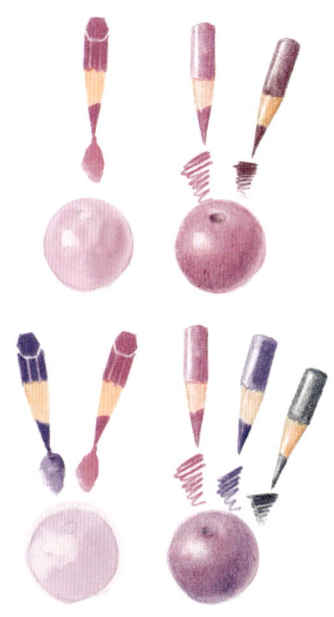

WATERCOLOR PENCIL: Middle Purple Pink #125

COLORED PENCIL SHADOW COLOR: Red Violet #194

LOCAL COLORED PENCIL: Middle Purple Pink #125

RED-VIOLET

WATERCOLOR PENCILS: Middle Purple Pink #125 and Purple Violet #136 (1:1)

COLORED PENCIL SHADOW COLOR: Dark Indigo #157

LOCAL COLORED PENCILS: Middle Purple Pink #125 and Purple Violet #136 (1:1)

VIOLET

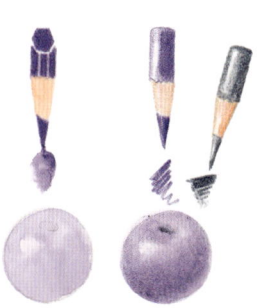

WATERCOLOR PENCIL: Purple Violet #136

COLORED PENCIL SHADOW COLOR: Dark Indigo #157

LOCAL COLORED PENCIL: Purple Violet #136

Violet is a secondary color created by mixing red and blue; however, for this exercise you can use a Purple Violet watercolor pencil. For additional practice, mix Middle Purple Pink #125 and Ultramarine #120 watercolor pencils (1:1), which are both bias toward violet, to create a bright violet color.

BLUE, BIAS TOWARD VIOLET

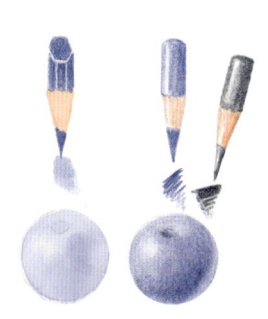

WATERCOLOR PENCIL: Ultramarine #120

COLORED PENCIL SHADOW COLOR: Dark Indigo #157

LOCAL COLORED PENCIL: Ultramarine #120

BLUE, BIAS TOWARD GREEN

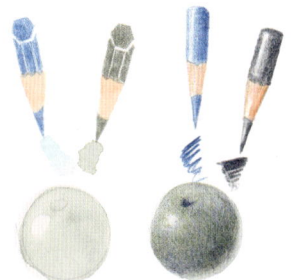

WATERCOLOR PENCILS: Ultramarine #120 and Earth Green #172 (1:1)

COLORED PENCIL SHADOW COLOR: Dark Indigo #157

LOCAL COLORED PENCIL: Cobalt Turquoise #153

CONTINUED

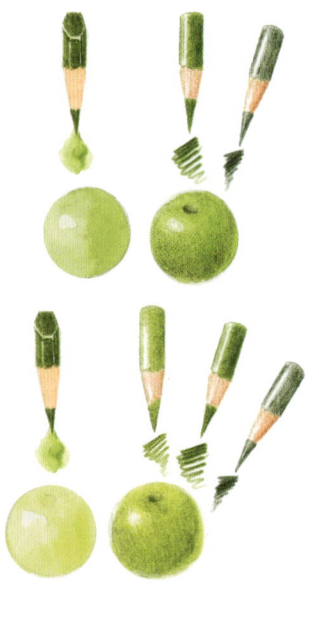

BLUE-GREEN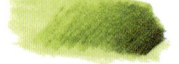

WATERCOLOR PENCIL: Permanent Green Olive #167

COLORED PENCIL SHADOW COLOR: Chrome Oxide Green #278

LOCAL COLORED PENCIL: Permanent Green Olive #167

GREEN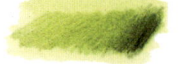

WATERCOLOR PENCIL: Permanent Green Olive #167

COLORED PENCIL SHADOW COLOR: Chrome Oxide Green #278

LOCAL COLORED PENCIL: Earth Green Yellowish #168

YELLOW, BIAS TOWARD GREEN

WATERCOLOR PENCILS: Permanent Green Olive #167 and Cadmium Yellow #107 (1:3)

COLORED PENCIL SHADOW COLOR: Earth Green #172

LOCAL COLORED PENCILS: Cadmium Yellow Lemon #205 and Earth Green Yellowish #168

Mixing Earth-Tone Colors and Shadows

In my sunflower color wheel, I have listed all my favorite colored pencils and shown when I use them in each of these petals. Use the following formula to help choose appropriate colors for your subjects:

LOCAL COLOR + SHADOW COLORS + HIGHLIGHT AND BURNISHING COLORS = SUBJECT COLOR

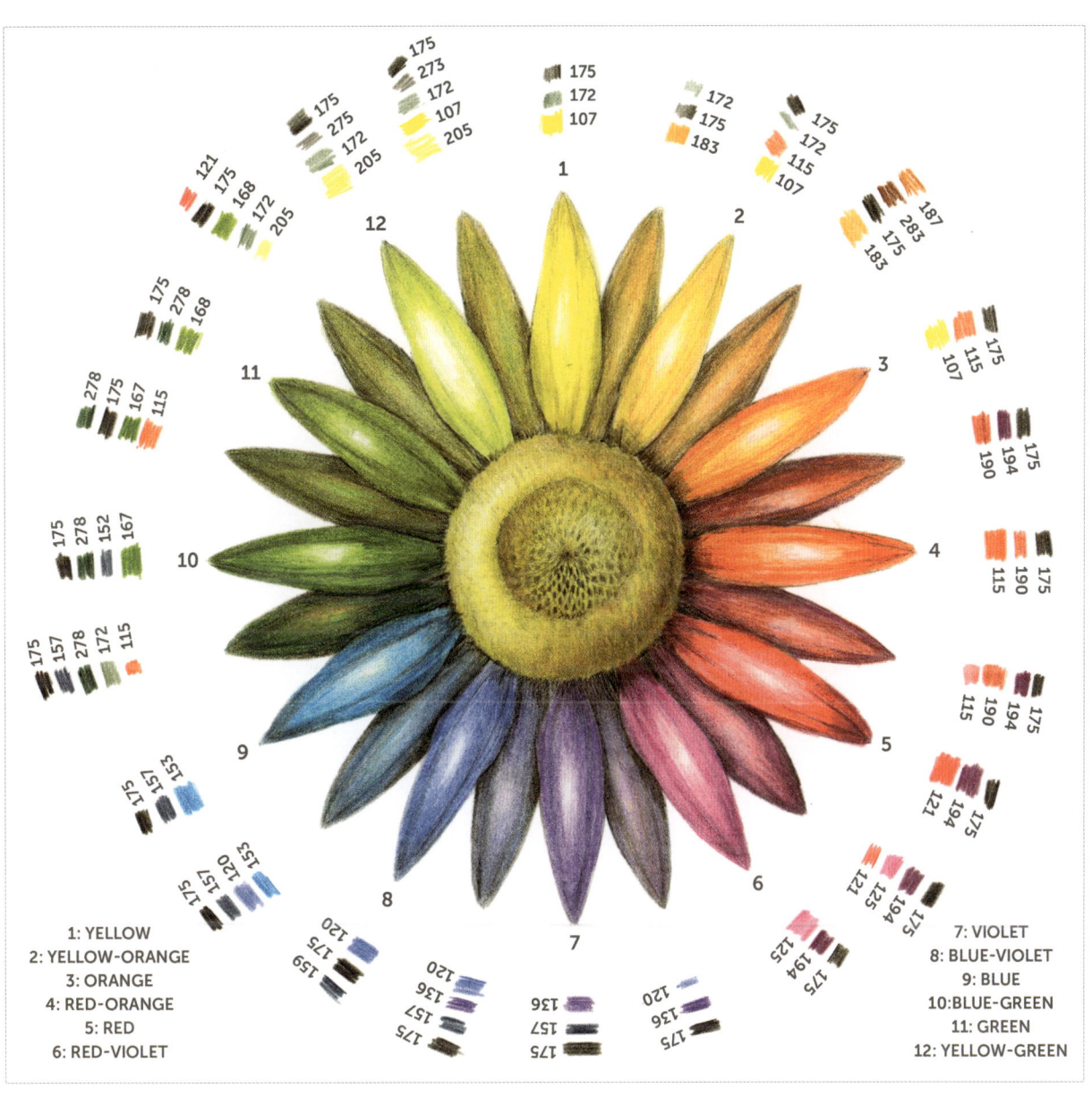

1: YELLOW
2: YELLOW-ORANGE
3: ORANGE
4: RED-ORANGE
5: RED
6: RED-VIOLET

7: VIOLET
8: BLUE-VIOLET
9: BLUE
10: BLUE-GREEN
11: GREEN
12: YELLOW-GREEN

CONTINUED

ADDITIONAL MATERIALS

- Watercolor brush #2
- Pale Geranium Lake colored pencil #121
- Cadmium Yellow colored pencil #107
- Ultramarine watercolor pencil #120
- Venetian Red colored pencil #190
- Red Violet colored pencil #194
- Dark Sepia colored pencil #175
- Ivory colored pencil #103

WATERCOLOR PENCILS

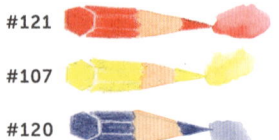

#121
#107
#120

COLORED PENCILS

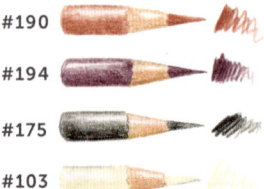

#190
#194
#175
#103

To mix an earth tone with either colored pencils or watercolor pencils, it's important to understand how to dull a color. Because we include important earth tones in our colored-pencil palette, you won't need to mix them; however, for a watercolor wash underneath an earth tone, you will need to do some mixing. In this lesson, you'll mix a wash that matches Venetian Red. There are an infinite number of ways to create this color, but let's first mix it with Pale Geranium Lake, Cadmium Yellow, and Ultramarine, three primary colors. The first step in understanding how to mix a color is to use words to describe it. The first questions to ask are, What is the hue? Is it closest to red? If so, most likely there will be more red in the blend you mix. If it's bright, it will most likely have only two primary colors present. If it is bias toward brown and dull, there will be some combination of all three primary colors present. Mixing complementary colors together is another way to create earth tones.

1. With a #2 brush, mix up a blend on your watercolor palette that's a small amount of Pale Geranium Lake with an even smaller amount of Cadmium Yellow in a 3:1 ratio. Make a swatch of this color on your paper. Next, add in a tiny amount of Ultramarine watercolor pencil, and make a swatch next to the first swatch. Notice the difference in the color. You have now added a tiny bit of the third primary, creating a duller color. **(A)**

2. Look at it next to a Venetian Red colored pencil and see if it is a lighter version of this color. If you need to adjust your color by adding more of a particular primary, go for it. Use your visual taste buds to feel if your color needs warmth (yellow), dulling (blue), or more red, for example. **(B)**

3. Draw your petal with a graphite pencil, and paint it with your Venetian red watercolor mix. Let it dry. **(C)**

4. With a Venetian Red colored pencil, use your grisaille toning on top of the petal. For shadows, add in Red Violet and a touch of Dark Sepia, and burnish with Ivory colored pencils. **(D)**

5. Reapply all colors on top for a smooth blending of colors. **(E)**

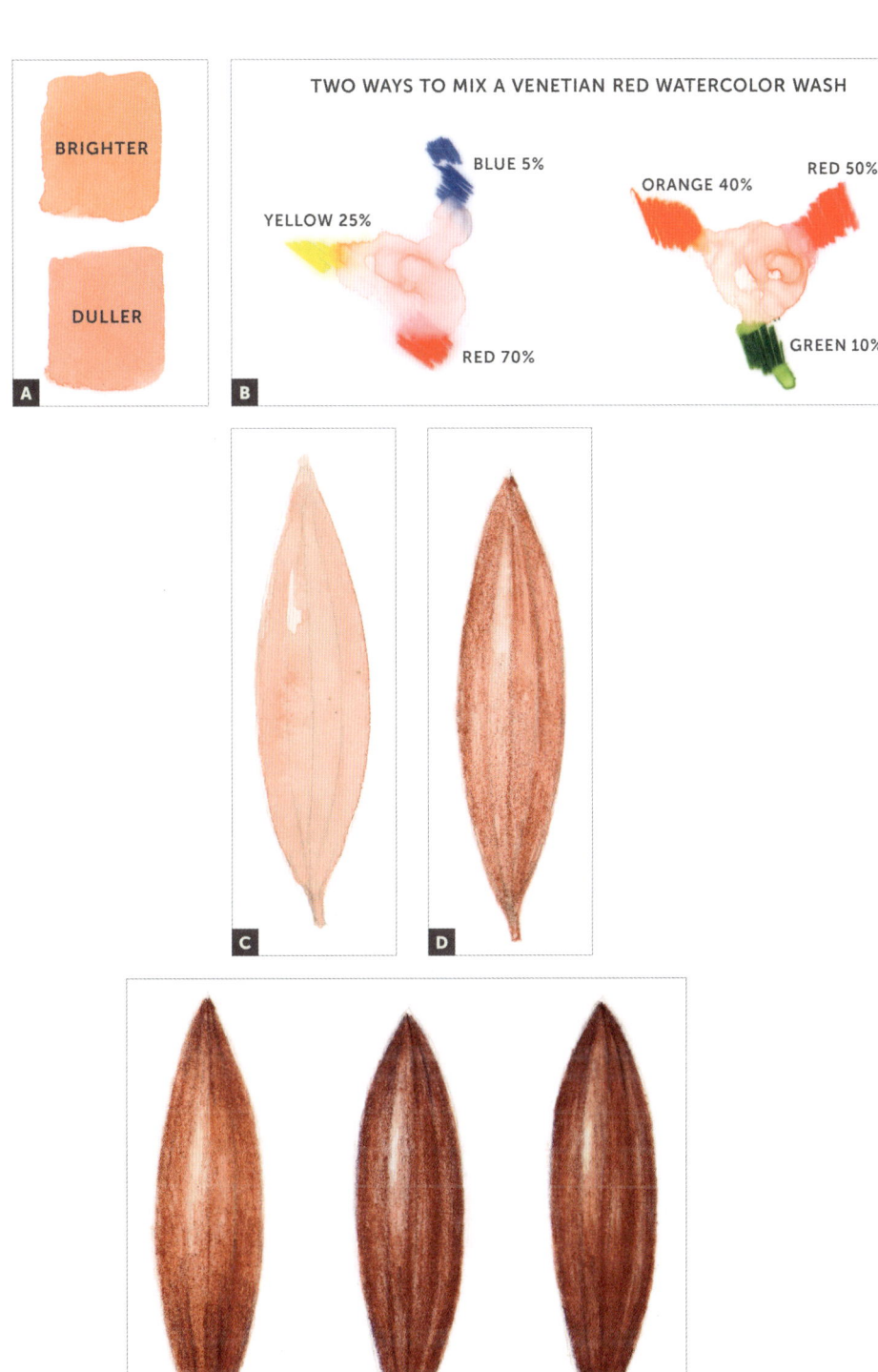

A — BRIGHTER / DULLER

B — TWO WAYS TO MIX A VENETIAN RED WATERCOLOR WASH

BLUE 5%
YELLOW 25%
RED 70%

ORANGE 40%
RED 50%
GREEN 10%

C

D

E

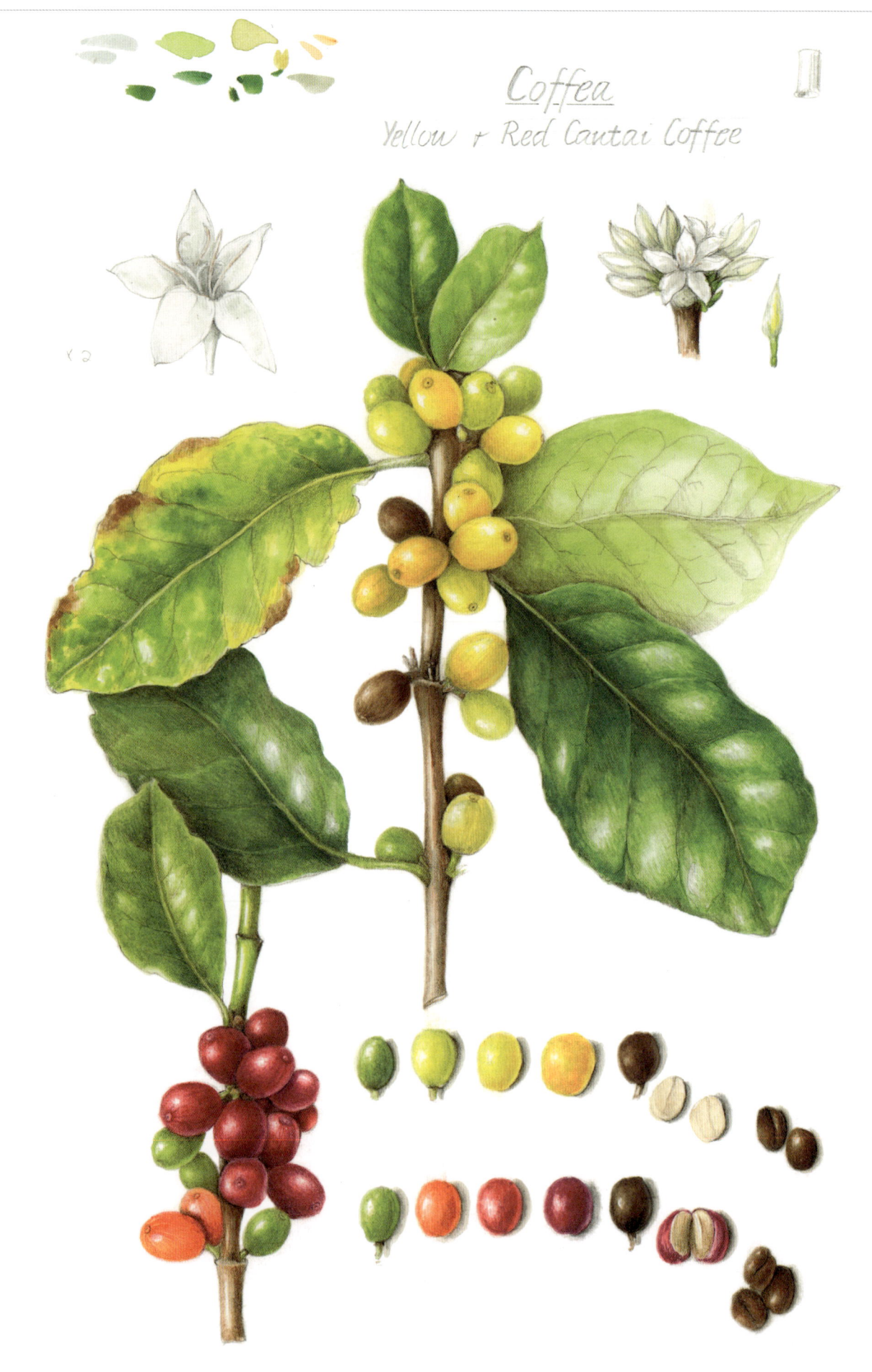

Coffea
Yellow & Red Cantai Coffee

CHAPTER FOUR

FOCUS ON LEAVES

Every plant has leaves, so learning to draw them is crucial to botanical drawing. There are many variations of leaves, and drawing them well is a lifelong pursuit. A description of a leaf's vein characteristics is an important part of botanical drawing as well as plant identification. Drawing subtle veining takes some practice and requires recognition of basic leaf-veining patterns. Plants fall into one of two leaf categories: monocot, a seed-bearing plant that has one cotyledon (the first leaf that appears on a seedling), and dicot, a seed-bearing plant that has two leaves that emerge first. It is helpful to know these two botany definitions that describe the two kinds of plants and their leaves.

Monocot leaves have parallel veining. They are usually long, strappy leaves that bend over, creating twists and turns. Parallel veining can be seen in grasses, tulips, irises, orchids, palms, corn, and bamboo. Dicot leaves have net veining, also known as branching veins. Roses, hibiscus, magnolias, and oaks have dicot leaves. Leaf drawing requires close-up examination of leaf veining, so a magnifying glass will be helpful. Because leaves are not very thick, we tend to think of them as being flat, but there are complex ways that light and shadow vary on leaves that can be used to make them look extremely dynamic and dimensional.

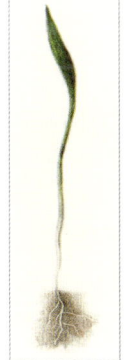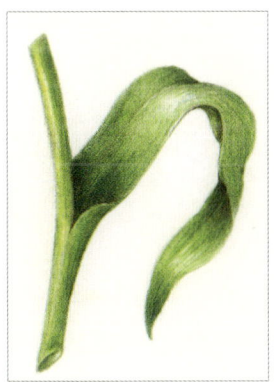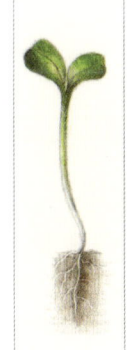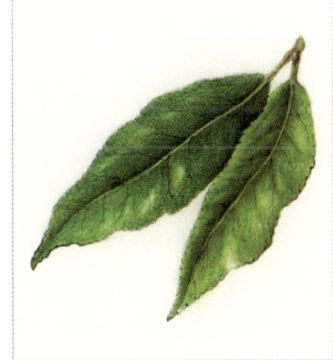

Monocot plants have one cotyledon and parallel veining.

Dicot plants have two cotyledons and net veining.

Parallel Veining and Overlapping Leaf

SUBJECT

Leaf with parallel veining, such as from a tulip, daylily, iris, grass, corn, or any other monocot plant

ADDITIONAL MATERIALS

- Chrome Oxide Green colored pencil #278
- Permanent Green Olive watercolor pencil #167
- Watercolor brush #2 or #6
- Permanent Green Olive colored pencil #167
- Other colored pencils to match the local color of your leaf

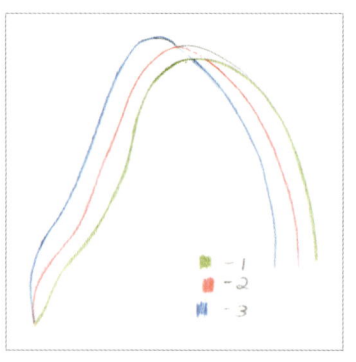

COLORED PENCILS

#278

#167

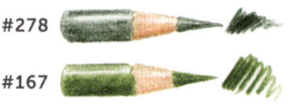

WATERCOLOR PENCILS

#167

Plants in the monocot family usually have parallel veining. These kinds of leaves all have similar characteristics, and drawing them offers a great way to practice the beginning concept of overlaps. An overlap is created when a leaf folds over, covering part of itself so that we don't see all of it. We can practice using shading to create contrast that contributes to a dynamic, dramatic, and dimensional drawing. Who doesn't get excited by the idea of showing a rolling and overlapping leaf with convincing drawing and toning! This lesson will take the mystery out of this process, but it will require practice to really understand the concept. Minor adjustments in a drawing will make a huge difference in your result, so be sure to draw lightly.

We'll also use a picture plane in this lesson, which is an imaginary perpendicular window right up against your subject. Taking measurements on this allows us to measure in perspective (see page 85).

As you work, observe and reference this drawing in which the closest edge is green, the center vein is red, and the outer edge is blue. It's important to keep track of all three lines; as when the leaf overlaps, an edge will disappear behind the rest of the leaf. We still want to visually continue all lines, even when we can't see them, so that they appear to connect. Repeat this lesson as many times as you like with different leaves that have parallel veining. Look at real leaves on plants to see how they twist and turn, and try to draw them realistically. You can also practice these overlapping techniques by cutting your own leaves out of paper and using tape to hold them in place.

1. Set up your leaf with a pleasing curve and overlap. You can tape your leaf down on paper to hold it in place, but keep the curve graceful.

2. Measure and draw the nearest edge of the leaf that you see, the edge right up to the picture plane. Make sure to draw a curving line that's graceful and not too angular. **(A)**

3. Next, draw the entire center curve of the leaf, describing the midvein in one continuous curve, even though part of the center will be obscured when the leaf rolls over. Draw the outside edge of the leaf; again, it will not be visible when it rolls over, but draw it lightly anyway, all the way through. **(B)**

4. Draw the width of the leaf as it is at the top, rolling over the side. Erase the lines that aren't visible. The secret to drawing a convincing rolling leaf is to have all the edges appear connected. **(C)**

5. Look at your leaf under a light source, or use your head light, and tone the overlapping area with a shadow color such as Chrome Oxide Green colored pencil to create a convincing bend with a good highlight on the top of the leaf. **(D)**

6. Using a #2 or #6 brush, add a layer of Permanent Green Olive watercolor that's darkest in the shadow area of the overlap and lightest at the highlight. **(E)**

7. Continue to add layers of colored pencil with Permanent Green Olive and other colors to match the local color of your subject, filling in details you see on your leaf, such as suggestions of the veining. **(F)**

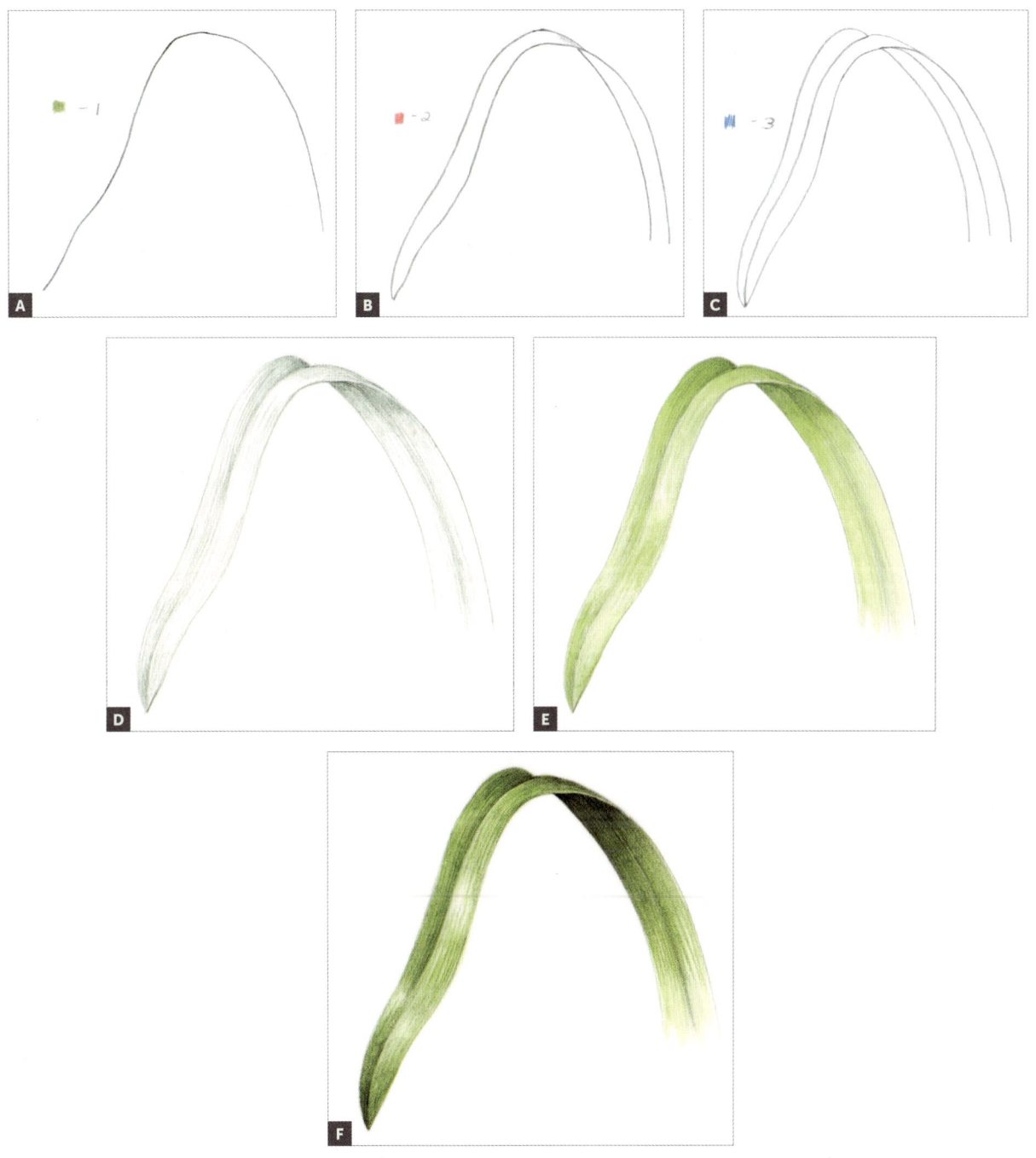

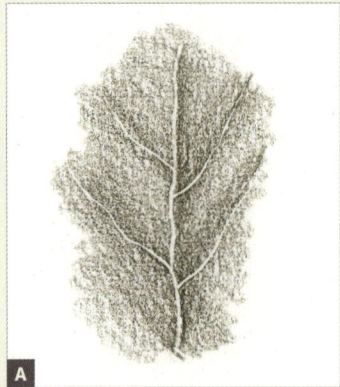

A

B

WATERCOLOR—

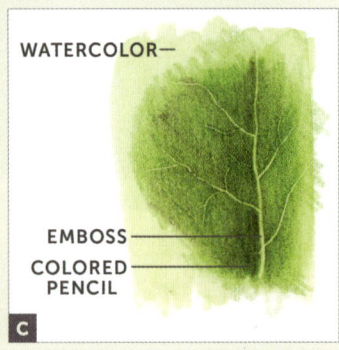

EMBOSS ————
COLORED ————
PENCIL

C

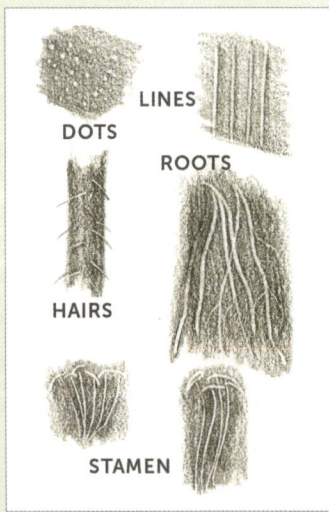

DOTS

LINES

ROOTS

HAIRS

STAMEN

EMBOSSING TECHNIQUE PRACTICE

When your botanical subjects have light, skinny areas, this is a good time to use an embossing tool. Thin roots, hairs, veins, or stamens are all potential subjects for embossing. Embossing tools are used in various crafts to create a recessed or indented pattern on a surface. On paper, they work to press fine lines into the paper surface. When you draw over an embossed area with a colored pencil, the embossed area stays light while the pencil tones cover the paper around the embossed area. If you paint over the embossed area with watercolor, the paper will be tinted, so keep this in mind when combining embossing techniques with watercolor. The trick to this technique is to mix embossing with regular drawing so that the embossing doesn't look too obvious. Practice keeping your embossing subtle and varied so that the lines are not stiff and regimented.

1. Take a moment to study your leaf subject. Be sure to use a magnifying glass so that you can see the pattern of the veins clearly.

2. To practice with the embossing tool, on a small section of your paper use a tool with a medium-size point to draw some veins. Draw with the tool as if it were a pencil, pressing it into the paper. Apply strong pressure at first, then ease up on the pressure, making the line thinner and thinner until it disappears. Practice this on your page to try different ways of using this tool. Draw over the embossed area with a light layer of Dark Sepia to see how the embossed lines remain the color of the paper and will show up once some toning is added around them. This is a great way to preserve thin, light areas on a drawing, such as veins of leaves. **(A)**

3. To practice tinting your veins, paint a small area of a green watercolor wash. **(B)**

4. Once dry, draw with the embossing tool over the watercolor and then apply Permanent Green Olive and Dark Sepia colored pencil, on top. Observe how the embossed areas are now tinted green, rather than the white of the paper. Veins are rarely as light as the paper, so tinting with a watercolor helps the veins look subtle. **(C)**

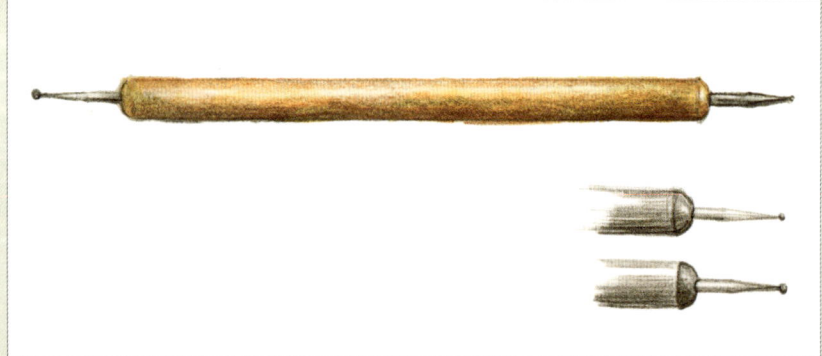

Basic Leaf-Veining Practice

This lesson explores the pattern of a leaf's net veining. This is any pattern with a strong center midvein and branching secondary and tertiary veins. You'll revisit leaves over and over again, and understanding their repeated patterns will be incredibly helpful.

Although you've probably looked at leaves thousands of times, as you work on this lesson really look closely at the leaf as if you are seeing it for the first time, and pay special attention to the veining. Veining is complicated because you're not only drawing the veins but also what appears around the veins on a leaf. Veins create recessed areas on the leaf's surface, which can create shadows and highlighted areas that are almost like pillows. I think of leaf veins as stitching on a quilt to understand their surface contour. If the veins are stitching, then what color do you use to draw them? Is it a light yellowish-green, or do you focus on the shadow created by the recessed area of the vein? It helps to think of the vein color as various shades of a color. Light and shadow make for variation in the veining, and your veins will look more realistic if you draw them this way. Sometimes veins are light, and sometimes they are dark. I constantly ask myself, "Am I drawing the vein, or the shadow created by the vein?" The answer is, I draw them both, sometimes on the same vein, and often I let my veins disappear as they approach the leaf margin. The veining looks different on the front and back of the leaf as well.

The key to realistic leaf veining is subtlety. Notice if the veins branch off before they reach the outer edges of the leaf. Make sure the veins aren't too straight or curving too much but have appropriate zigzagging to look realistic. Observe and include the proper width variation and tapering at the tips and edges. For good toning on the surface contour of a leaf, remember how the light source reveals the distinct planes of the leaf. As you draw your leaf, keep in mind that the midvein is often the widest and lightest in color. The secondary veins will be thinner and slightly darker in color, closer to the color of the body of the leaf, and the tertiary veins will be even skinnier and closer in color to the body of the leaf. Observe these veins under magnification to see what I'm describing.

To enlarge the leaf pattern from your subject and exaggerate the veining, simply measure the width of your midvein and multiply it by 2, making it twice as wide (roughly 1/8 inch wide). But note that it will taper and get thinner as it moves toward the top of the leaf.

CONTINUED

CONTINUED

SUBJECT

Leaf with net veining, such as from baby kale, a rose, or a hibiscus

ADDITIONAL MATERIALS

- Dark Sepia colored pencil #175
- Watercolor brush #6
- Permanent Green Olive watercolor pencil #167
- Permanent Green Olive colored pencil #167
- Embossing tools in several sizes
- Wax paper (optional)
- Other green colored pencils to match your leaf's local color
- Black Verithin pencil (optional)

COLORED PENCILS

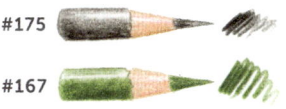

#175

#167

WATERCOLOR PENCILS

#167

1. Lightly, with graphite pencil, draw a small enlarged section of the midvein and secondary veining on a baby kale leaf. Draw it slightly wiggling or irregular, as it is neither straight nor curved. It most closely resembles a series of straight lines that keep moving slightly in different directions. Lightly indicate where the secondary veins will appear, making them irregular as well.

2. Use a Dark Sepia colored pencil to add a light layer of tone next to the midvein on the left and right side. Add a second layer of tone on the left plane of the leaf next to the midvein and also right above each secondary vein. **(A)**

3. Using a #6 brush, paint a light layer of Permanent Green Olive watercolor on top of the leaf. You can graduate the watercolor a bit by starting darker on the left side of the leaf closest to the midvein and getting lighter toward the outer left side of the leaf edge. On the right side of the leaf, graduate the watercolor from a little darker at the leaf edge and lighter toward the midvein. This reinforces the idea that the two sides of a leaf receive light differently because they are two separate planes, similar to an open book (see page 78).

4. With a Permanent Green Olive colored pencil, grisaille tone over the watercolor, starting on the left side of the midvein. Begin at the edge of a secondary vein and tone toward the next secondary vein, but stop toning just before you reach the next vein, creating a slightly lighter skinny line or secondary vein. Add grisaille toning next to each secondary vein and toward the center of each area, leaving it slightly lighter in the center, to create a slight pillowing effect between the secondary veins. You can sometimes observe this pillowy look on leaves in nature. **(B)**

5. Add tone on the right side of the leaf, but this time tone darker at the right edge and leave bigger highlights in each section. Also tone a bit next to the center midvein.

6. Create some tertiary veins very subtly with the embossing tool; either do this directly on top of your drawing or try putting down a piece of wax paper and embossing over it, which will remove a bit of the pencil. Remove the wax paper and tone over the area with colored pencil—the vein will show up more but will be subtler than the more prominent veins. **(C)**

7. Finish with more Dark Sepia and green colored pencils. If your veins are too light, darken them a bit so that they're subtler. You can darken inside an embossed area with watercolor or with a very sharp pencil (Verithin pencils work well for this purpose).

8. Turn over your leaf and study the pattern on the back; notice that the veins are raised, making them appear more prominent. The midvein can be toned like a cylinder with a cast shadow to make it appear to sit on top of the leaf surface. Draw a section of the midvein on the back of the leaf with Dark Sepia to practice this concept. **(D)**

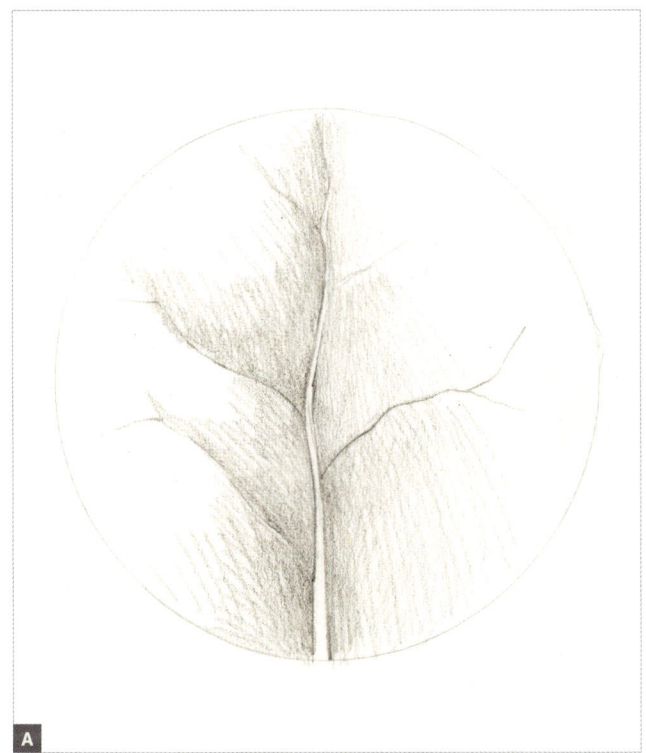

A

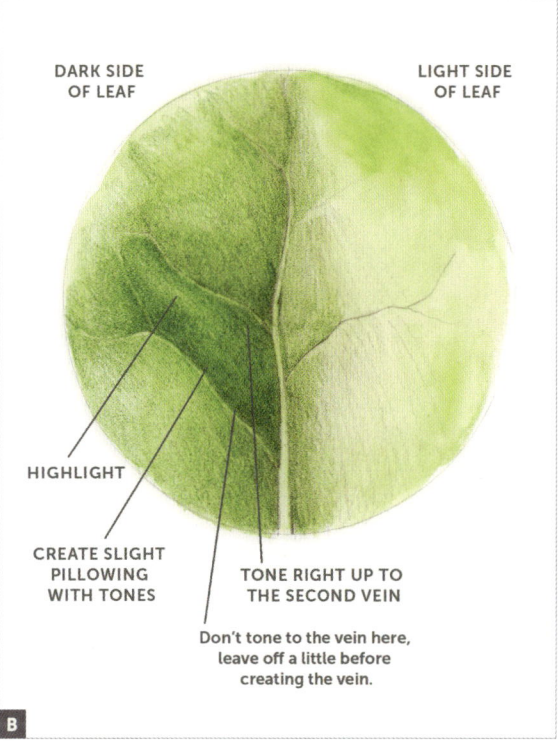

DARK SIDE
OF LEAF

LIGHT SIDE
OF LEAF

HIGHLIGHT

CREATE SLIGHT
PILLOWING
WITH TONES

TONE RIGHT UP TO
THE SECOND VEIN

Don't tone to the vein here,
leave off a little before
creating the vein.

B

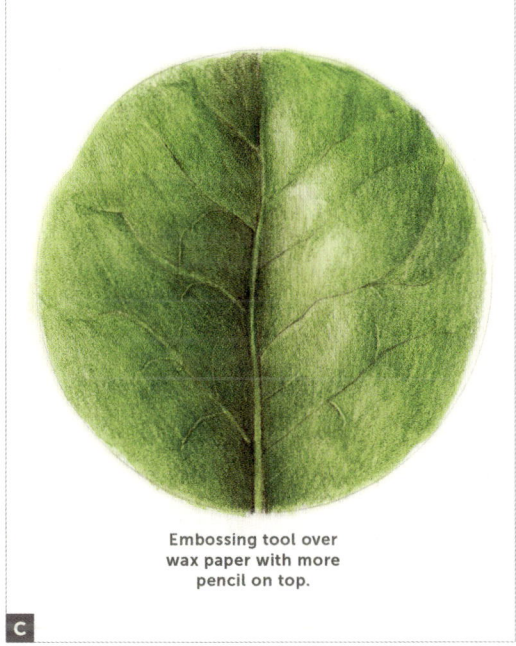

Embossing tool over
wax paper with more
pencil on top.

C

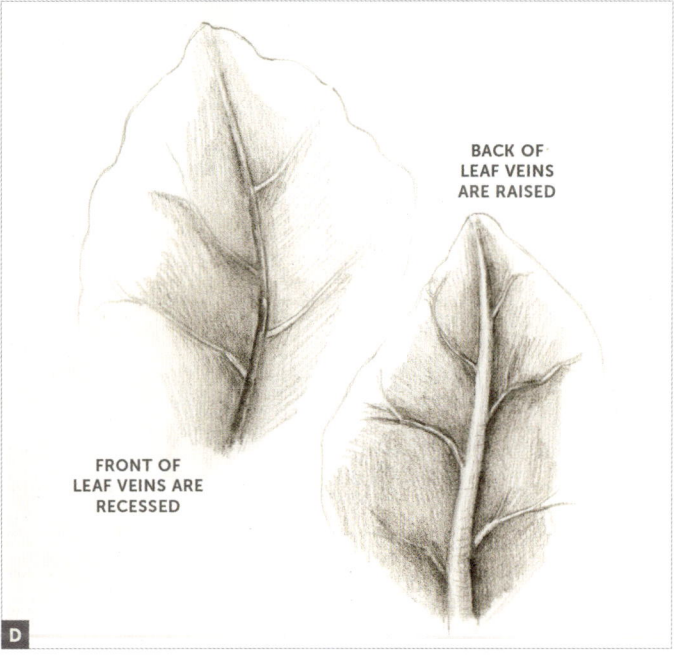

BACK OF
LEAF VEINS
ARE RAISED

FRONT OF
LEAF VEINS ARE
RECESSED

D

A Basic Leaf with Net Veining

SUBJECT

Leaf with net veining, such as a baby kale leaf (or any leaf that has a strong center midvein with branching secondary and tertiary veins)

ADDITIONAL MATERIALS

- Dark Sepia colored pencil #175
- Permanent Green Olive watercolor pencil #167
- Watercolor brush #2 or #6
- Permanent Green Olive colored pencil #167
- Embossing tool (small)

COLORED PENCILS

#175
#167

WATERCOLOR PENCILS

#167

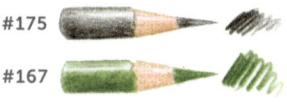

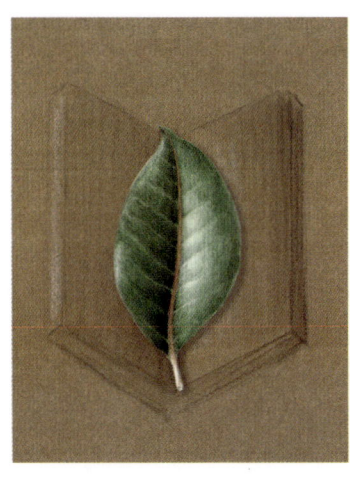

Leaves with net veining have a center midrib or vein creating two distinct planes that are affected differently by light. Think of the planes of a leaf as two sides of an open book. When light hits these two planes, one side receives more light and the other side more shadow. If you tone your leaf using this concept, it will look dimensional even though a leaf has very little thickness.

Once you understand the veining pattern, you can draw veins as a suggestion and not make them too prominent. Leaf tones in a finished drawing will be close in value, and veins will be subtle. It is a delicate balance to tone a leaf correctly. In the end, the veining should be close in value to the rest of the leaf so as not to have veins that are stiff and overpowering.

1. With a graphite pencil, lightly draw your leaf life-size. Begin your drawing with the center vein and then draw the outside edges of the leaf. With a Dark Sepia colored pencil, use tones to depict the leaf planes, then slowly add more and more detail indicating the secondary veins. Tone a light layer of Dark Sepia to get the leaf started, but don't make it too dark at the beginning. **(A)**

2. Tone the left side of the leaf darker. Shade from dark to light, starting at the midvein and toning lighter toward the left edge of the leaf. Add a little bit of tone next to the midvein on the right side of the leaf by leaving a light vein and then toning a bit toward the right side of the leaf.

3. Add tone to the right leaf edge, and tone lightly toward the center of the leaf along each secondary vein. This side of the leaf should not be as dark as the left side. Veins are thicker or wider at the base and always get thinner toward the tip of the leaf. Because secondary veins usually branch out and appear to disappear before the outside edges, be sure to make them smaller and less defined. **(B)**

4. Using Permanent Green Olive watercolor pencil and a #2 or #6 brush, add a layer of watercolor wash over the Dark Sepia pencil, gradating the wash to indicate the two sides of the leaf and making the left side darker. Paint a very light layer of green wash over the center vein as well. **(C)**

5. To add subtle veining and the slight pillowing between the veins, layer colored pencils in Permanent Green Olive and more Dark Sepia. Remember to keep the veining subtle, irregular, and not too straight. Add a bit of embossing for the secondary veins. Remember that veins of this kind don't usually go all the way out to the leaf edges. They tend to branch off before the edges and get thinner and thinner until they disappear. **(D)**

6. Look at your leaf and decide if you've gotten too literal with your veining. This happens often in the beginning. If you feel your veins look kind of like fish bones rather than subtle veins, don't feel discouraged. Narrow the veins a bit, and make the color values closer in value by adding more watercolor on the veins. Draw another leaf, and try not to focus on the veins at all but keep them really, really subtle. **(E)**

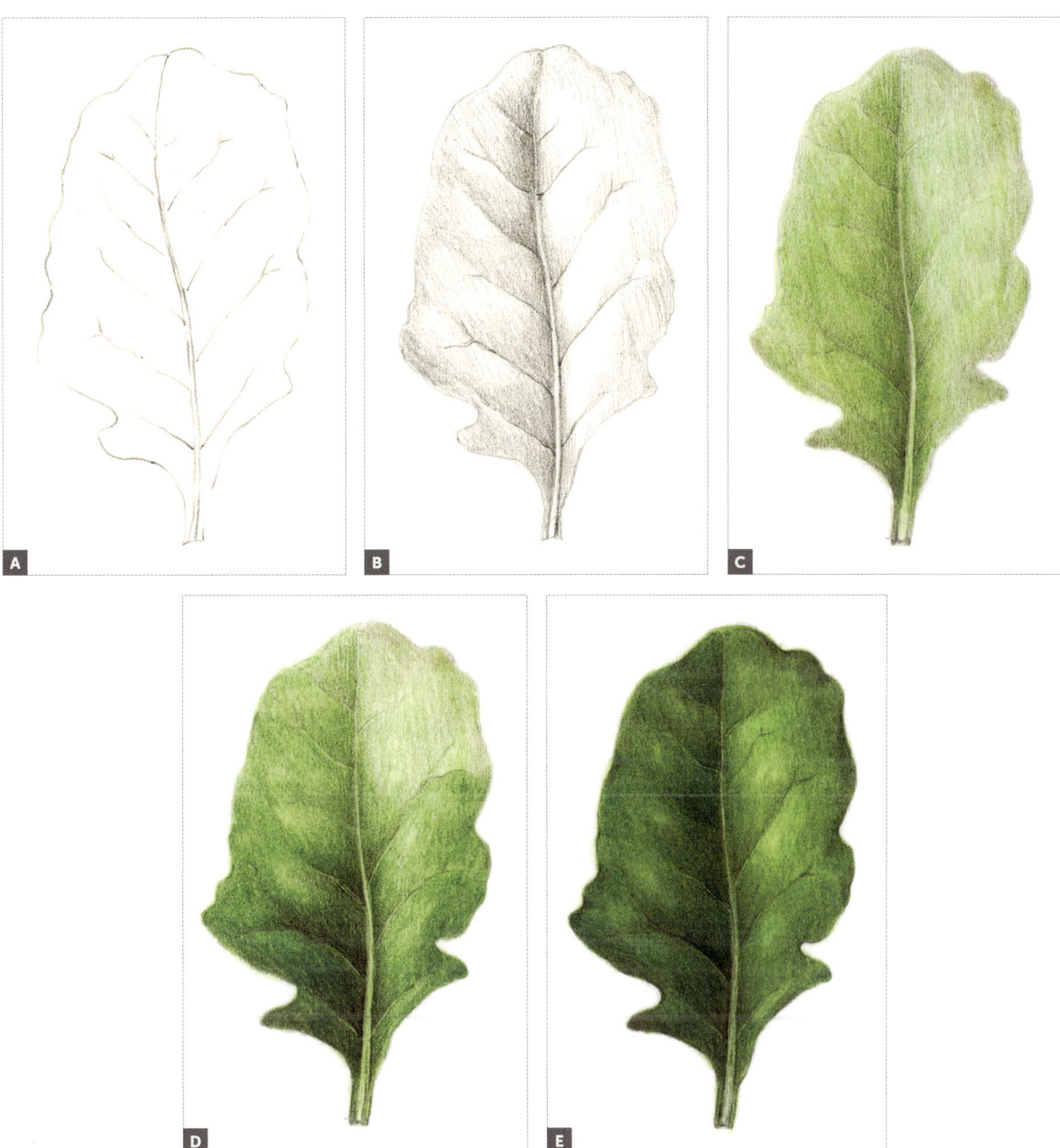

Colorful Leaf with Watercolor

SUBJECT

Colorful leaf

ADDITIONAL MATERIALS

- Watercolor brush #6
- Watercolor and colored pencils to match the colors of your leaf
- Embossing tool (optional)
- Warm Grey IV watercolor pencil #273 (optional)
- Cool Grey 70% Verithin pencil (optional)
- Black Verithin pencil (optional)

WATERCOLOR PENCILS

#273

This lesson uses watercolor washes in a loose style, working with the wet-on-wet technique to create vibrant colors, which is so much fun. After your watercolor dries, the details of the leaf veining are created with colored pencil. Collect some colorful fall leaves to use as your models. These instructions show my process, but rather than copying my leaf, use your own leaves for this lesson. When I suggest a specific color to use, keep in mind it will most likely be a different color for your leaf, but the process remains the same.

1. Draw a light outline of your leaf with graphite pencil. **(A)**

2. Wet the paper inside your leaf drawing with a large brush. Add a layer of yellow watercolor and then, while the paper is still wet, add some orange watercolor at the top edge and some brown watercolor where it appears on the leaf, allowing the colors to bleed into each other. **(B)**

3. Once that layer is dry, add subtle veining with colored pencils. Use some subtle embossing on veins if you think it works on your leaf, but always look at your leaf for a guide.

4. Add more watercolor if needed, wetting the paper again if it has dried so that the watercolor spreads rather than creating flat edges when applied. **(C)**

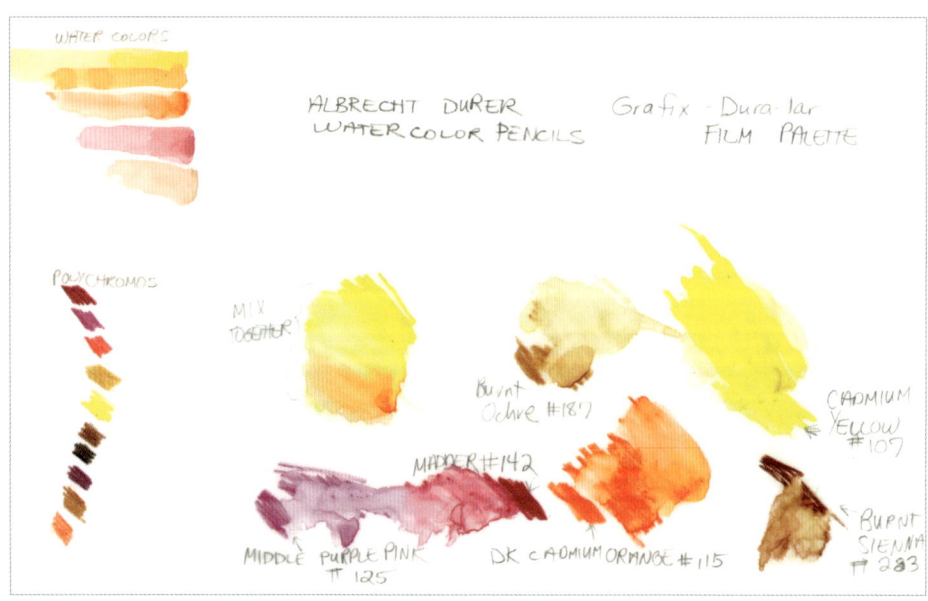

5. When dry, apply grisaille toning layers of colored pencil to saturate the colors—look at your real leaf as a guide. Emphasize veining and details, but keep them subtle. Make one side of the leaf slightly darker to show that it's under a good light source that creates two distinct planes. **(D)**

6. If desired, add in a cast shadow. Start with a light layer of Warm Grey IV watercolor, and keep the outer edge of the shadow gradually fading into the background. When the watercolor is dry, layer Cool Grey 70% Verithin pencil and Black Verithin right where the shadow touches the leaf edge. **(E)**

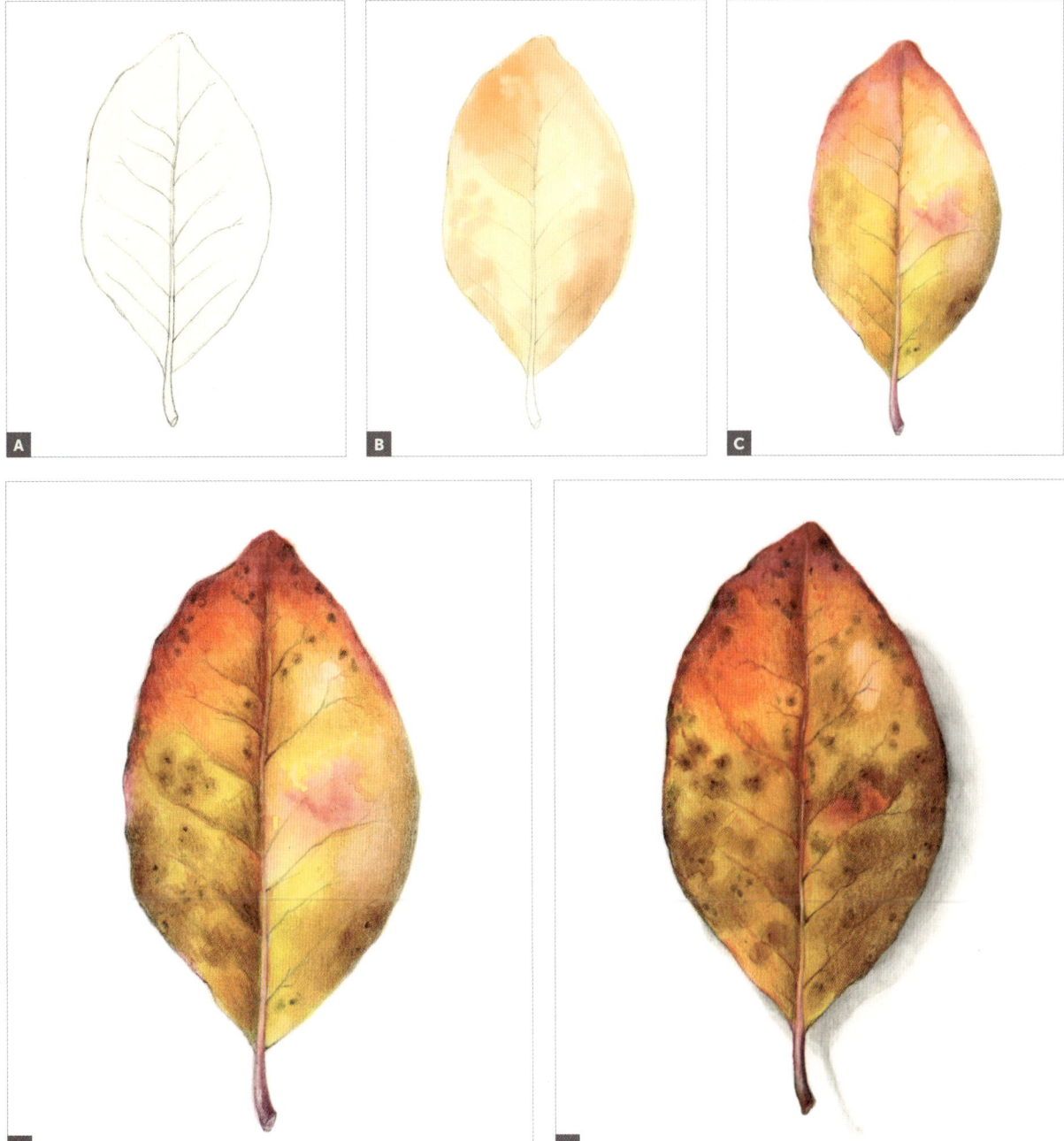

Leaf Front and Back

SUBJECT

Leaf with net veining, preferably one with a distinct front and back

ADDITIONAL MATERIALS

- Dark Sepia colored pencil #175
- Watercolor brush #2 or #6
- Watercolor pencils to match the local color of your leaf
- Earth Green colored pencil #172
- Burnt Sienna colored pencil #283

COLORED PENCILS

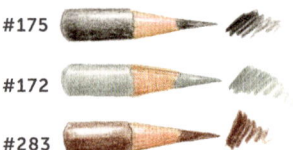

#175

#172

#283

Choose a leaf with net veining. I chose a *Magnolia grandiflora* leaf, as they are thick and leathery, and keep their shape for a long time. The planes of this leaf are very distinct, and the front and back are quite different in color. Draw both sides of the leaf next to each other. It's fun to see both sides of the leaf for comparison.

Study the back of the leaf. The midvein is more pronounced and can be likened to a cylinder on top of the leaf. As such, it can be toned like a cylinder. Right next to this prominent midvein will be a slight cast shadow, which will help raise the midvein on the leaf's surface. Compare it to the front of the leaf, where the veins are sitting inside the leaf and are recessed. The other difference between front and back is that the back of the leaf is like the outside of a book, so it will receive light differently than the front of the leaf.

1. With a graphite pencil, draw a light outline of the front of the leaf, and lightly indicate the center vein and secondary veining. Notice the angle of the secondary veins as they attach to the midvein and that these veins don't usually continue to the edge of the leaf. Turn the leaf over and draw the back. **(A)**

2. Apply Dark Sepia toning to the front of the leaf, using the same light source and grisaille toning as the lesson on the basic leaf with net veining (see page 78).

3. Repeat the Dark Sepia toning on the back. In addition to the obvious color difference between the front and back of this leaf, there are other characteristics that help distinguish the front from back: the back of the leaf will be toned in almost the opposite manner of the front. Plus, it is darker on the right side of, and has a stronger shadow cast by, the prominent midvein. Try to describe these characteristics with your Dark Sepia pencil grisaille toning. **(B)**

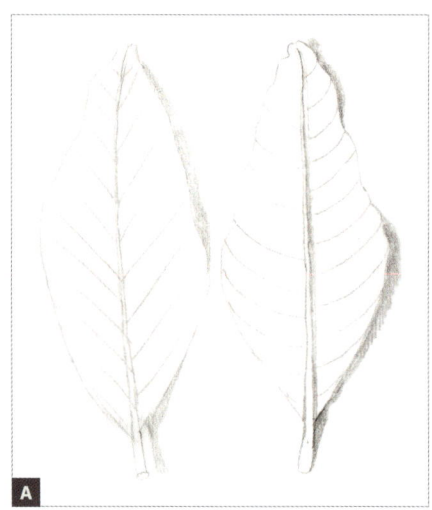

A

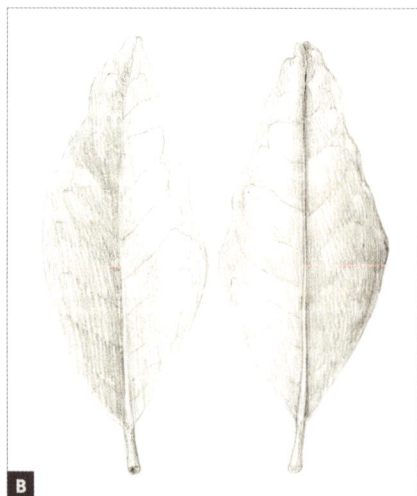

B

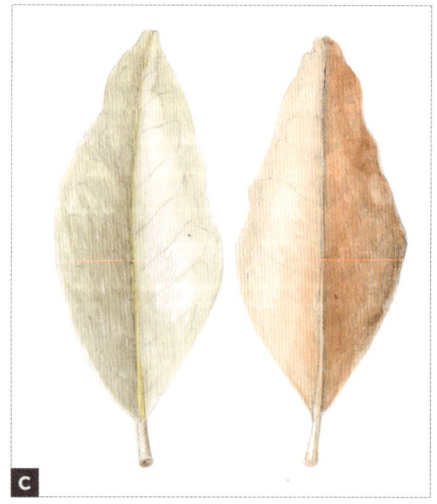

C

4. Using a #2 or #6 brush, add a layer of graduated watercolor to each leaf drawing in the appropriate base colors. **(C)**

5. Start to indicate more color and veining with Dark Sepia, Earth Green, and Burnt Sienna colored pencils. **(D)**

6. Continue to layer colored pencils, and try to keep the veining subtle and the light source well described on the two distinct sides of the leaf.

7. Add in subtle tertiary veins and a cast shadow on each leaf, which will help describe the ways the front and back of the leaf sit on a surface. **(E)**

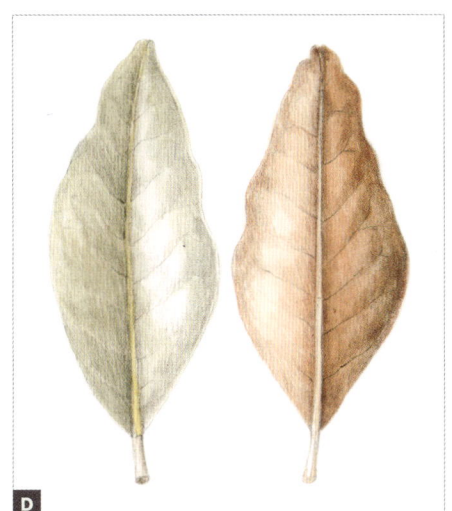

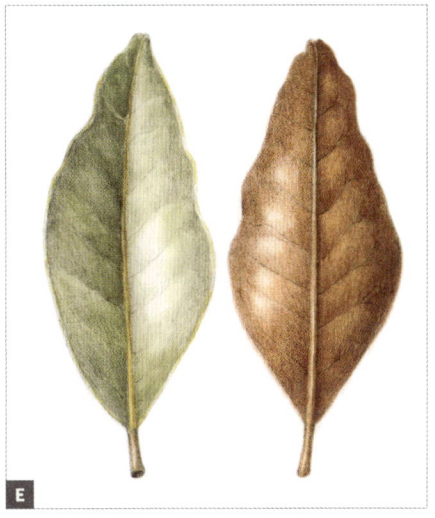

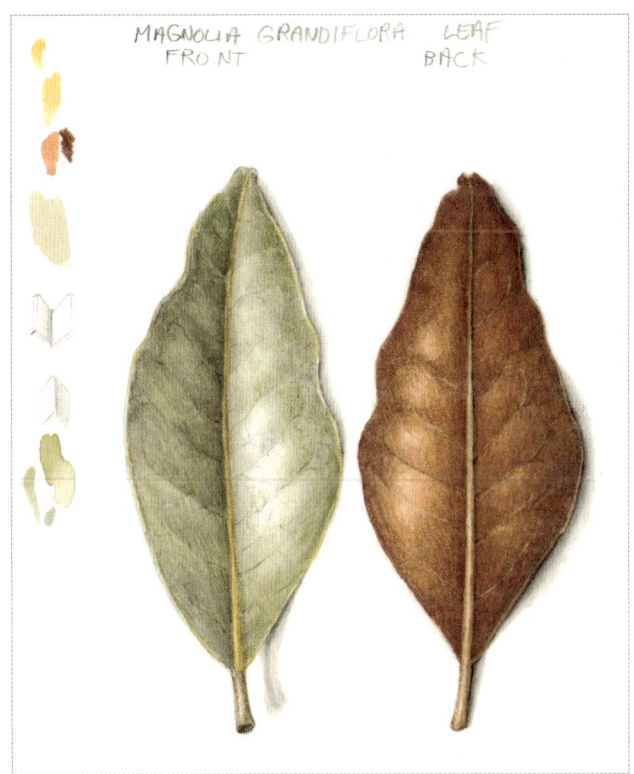

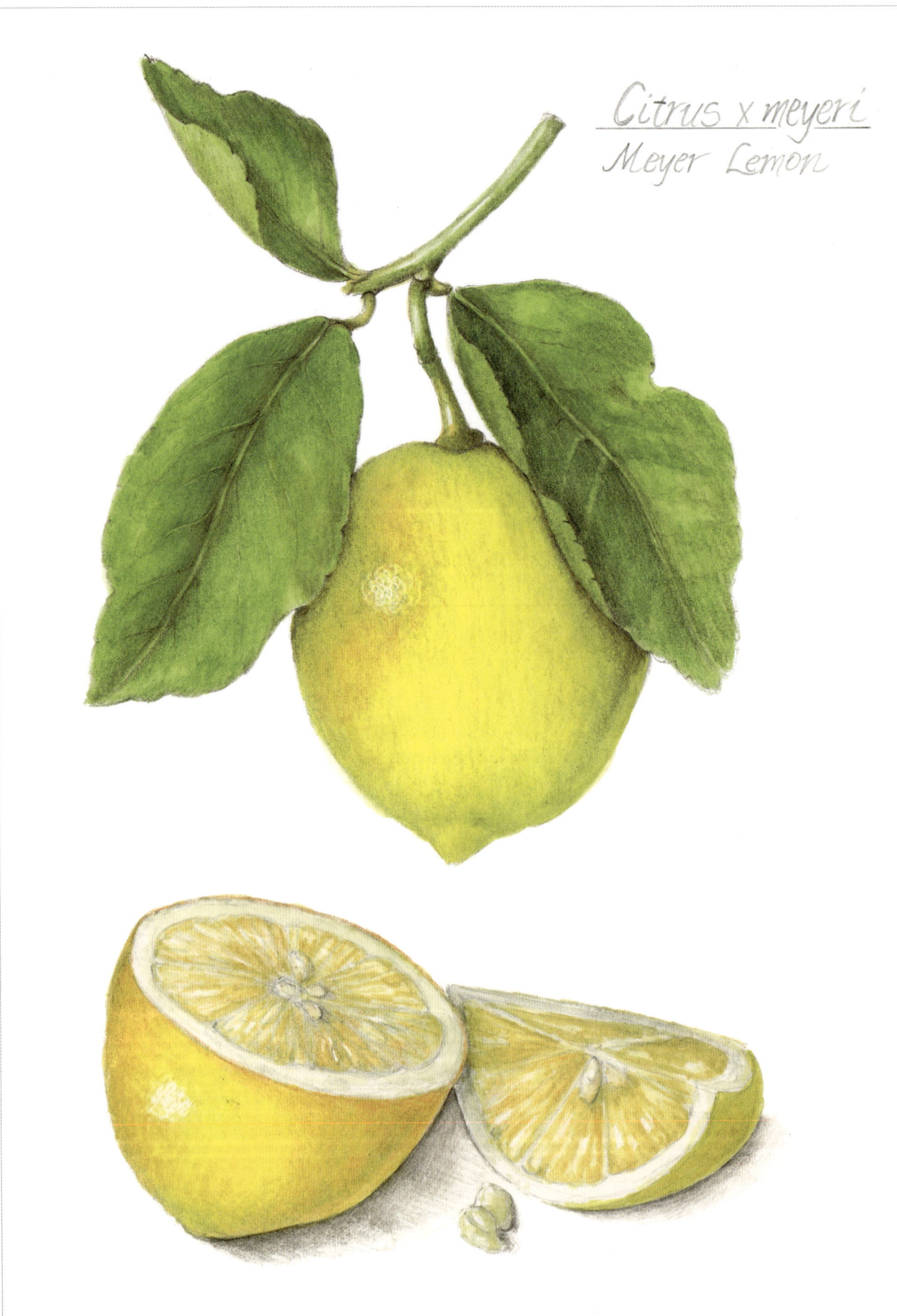

Citrus x meyeri
Meyer Lemon

CHAPTER FIVE

PERSPECTIVE AND MEASURING ACCURATELY

There are four basic components artists use to create the illusion of three-dimensional space on a two-dimensional piece of paper: (1) shadows and light, (2) overlaps, (3) perspective, and (4) color variation. We've been practicing using shadows and light to make a form look three-dimensional with natural color, and we use the concept of overlaps throughout this book. This chapter focuses on using perspective to make the form and space look three-dimensional. To do this, we'll create the illusion of depth using foreshortening, which is when a part of a subject's depth is diminished. Understanding foreshortening is crucial, as is choosing a particular view of a subject. While there are many aspects to the larger concept of perspective, for our botanical purposes, we'll focus on using circles that become ellipses. The position of the viewer's eye in relation to an object can change, which creates various foreshortened views. An easy way to start visualizing this is to do the following exercise using a cup.

Imagine that there's a window (or what we refer to as the picture plane) right up against your cup. I've created a grid on tracing paper that I can place in front of my drawing to help with measuring. I hold my ruler at a 90-degree angle and make sure it actually touches my subject up against the imaginary window. Because the window is flat, in effect you've transformed the three dimensions of the cup into a two-dimensional view. If you measure the height, width, and depth this way, you'll have a view of the cup in perspective. You can also take a photo to help practice foreshortening, because a photo transforms the three-dimensional world into two dimensions for you. I caution you, though, to not rely on photos for drawing structure and details in your plant, but only as a reference and a guide to help with perspective. Photos can help, but they are also misleading, so please focus on drawing from real three-dimensional subjects.

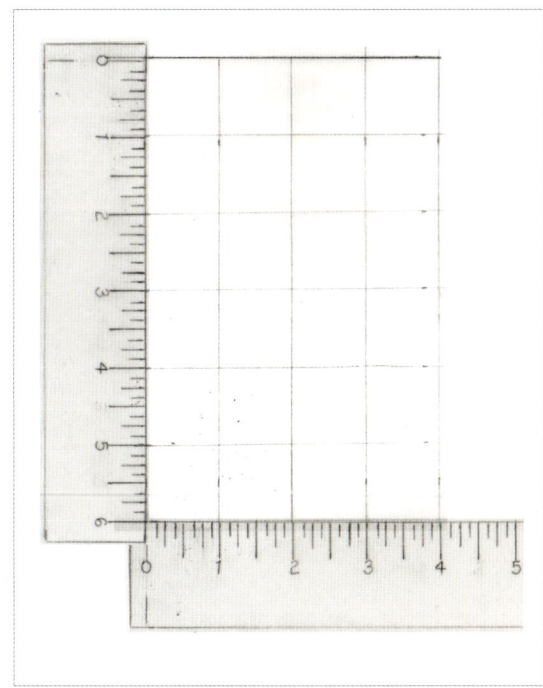

Perspective on a Cup

SUBJECT

Cup or small drinking glass

To measure correctly, I use a clear ruler that I can hold right up against the imaginary picture plane, right up against my subject. This may seem cumbersome and uncreative, but I find it to be liberating and freeing to check my measurements so that my proportions and sizes are accurate. This way, I don't have to continually fix and remeasure my drawing when something doesn't feel right. With a few quick measurements up against the picture plane at the beginning of a drawing, you'll save lots of time down the road. Nature's proportions are so lovely, and they often describe the difference between one plant and another, so getting them right is important. In this lesson, we will do some simple measuring and drawing perspectives of a cup on scrap paper. Once you've practiced visualizing and drawing the different ways of seeing a foreshortened view, use these simple drawings of your glass to help you when drawing with real plants.

1. Hold a cup up so that you are looking directly inside it. Measure the height and width of the top rim of the cup. Notice that this is a circle that has an equal height and width. **(A)**

2. Start to tilt the cup away from you just a bit. Measure the height and width of the rim again. Notice the width remains the same, but the height is diminished, or foreshortened. You are also beginning to see the outside of the cup. Draw this view. **(B)**

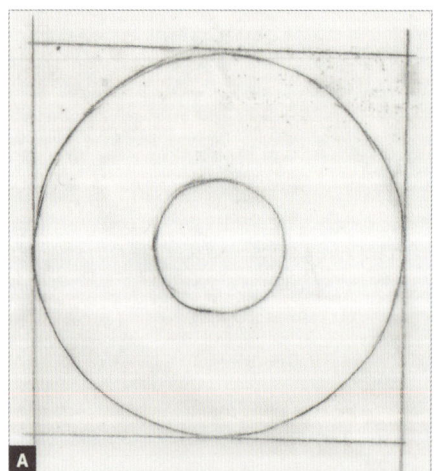
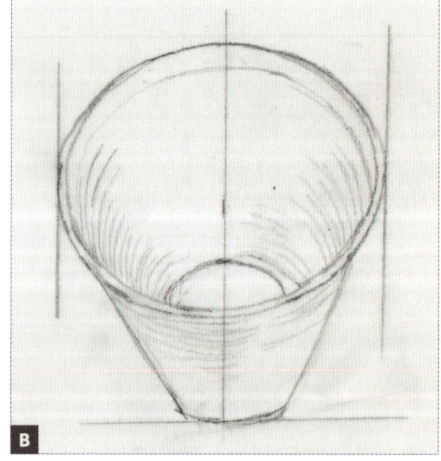

3. Tilt the cup even farther from you, and notice the rim is now an even more foreshortened ellipse and that you can see more of the outside of the cup. Draw this view. **(C)**

4. Hold the cup at eye level. Now the rim of the cup has become a straight line and you see the entire front of the cup. Draw this view. **(D)**

5. Rather than holding the cup straight, tilt the axis to about 45 degrees. You will now have a cup in perspective but at an angle, which is useful for creating varied and pleasing compositions, especially with flowers. Draw this view. **(E)**

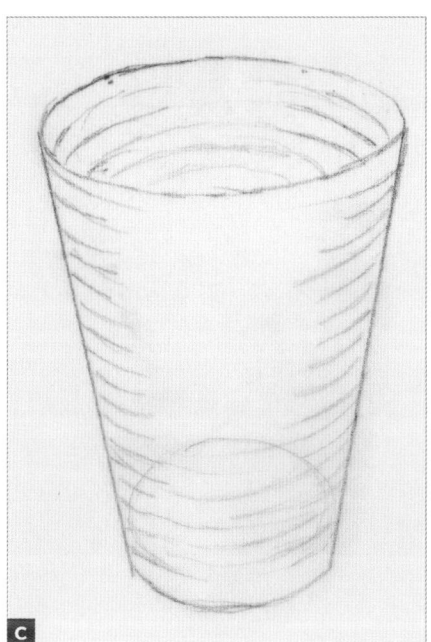

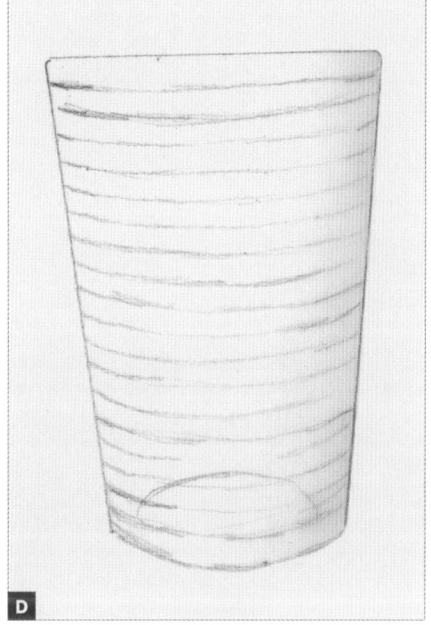

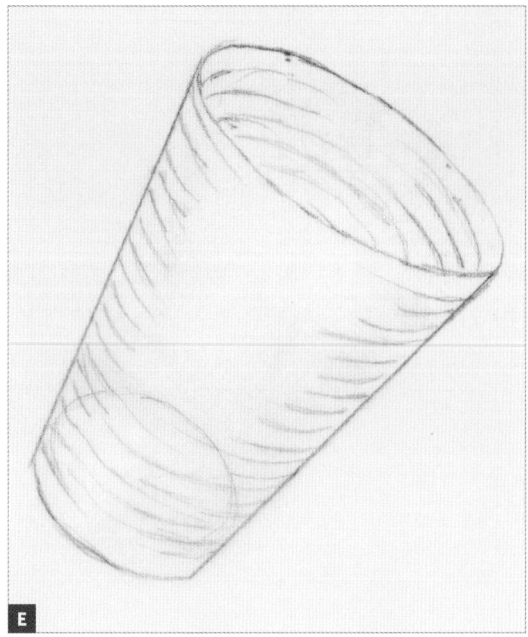

Measuring a Cross Section of a Fruit in Perspective

SUBJECT

Lemon or other fruit

ADDITIONAL MATERIALS

- Tracing paper
- Colored pencils to match your fruit's local color

Cross sections make great additions to botanical compositions. A cross section of a fruit forms a round or elliptical shape, which makes it perfect for practicing perspective measuring techniques. In this drawing of a lemon, refer back to the cup drawings in the previous lesson to help visualize the concept.

To draw an ellipse that's rounded and doesn't have sharp edges, it's important to draw each quadrant of the ellipse separately, turning your paper as needed so that you're relaxed when drawing. Using the diagram below as a reference, first draw a vertical height and width line **(1)**, then draw each quadrant of the ellipse. Start by drawing each apex with a curve **(2)**. Then continue to complete each quadrant to draw a continuous curve without any sharp angles **(3)**. Avoid having your ellipse look like the shape of an eye.

Repeat this lesson as many times as you like, changing your view and with different subjects. I recommend trying this with any fruits or vegetables. Use the techniques for adding color that you've practiced to complete these drawings, and add details such as seeds and other cut fruit sections to your compositions.

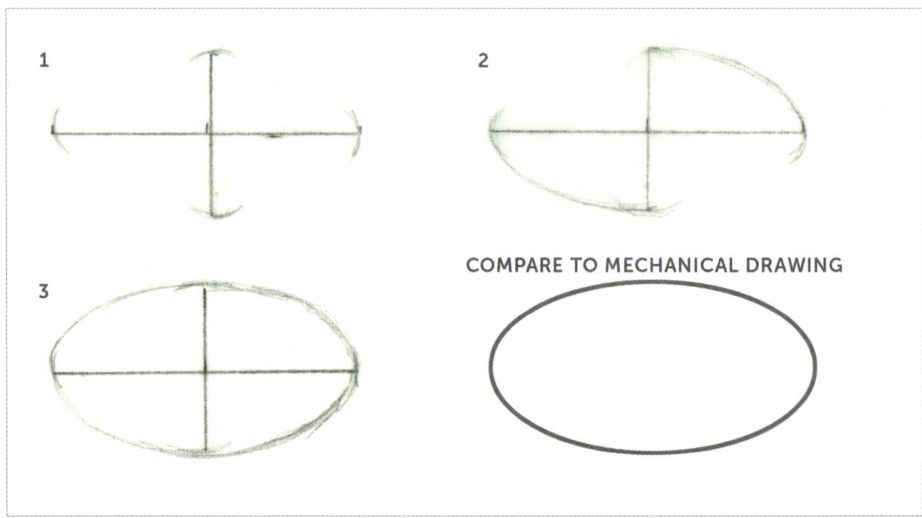

1

2

3

COMPARE TO MECHANICAL DRAWING

1. Cut your fruit in half. You now have a cross section to draw.

2. Set up the cross section of your fruit using a foreshortened view and measure the ellipse created, remembering to use your imaginary picture plane for placement of your ruler.

3. On tracing paper, start the drawing with a center axis (an imaginary line in the center of a subject usually emanating out of the stem, where it connects to the fruit or flower) first. Measure and draw the height, width, and depth of the cross section ellipse and the form of the body of the fruit behind it, working lightly with a graphite pencil. **(A)**

4. Practice drawing elliptical cross sections on tracing paper. Once you feel confident, draw in the details of your fruit cross section on good paper. Notice how the sections of a citrus fruit all radiate out from the center axis. Follow the steps for drawing a subject in color (see page 59) to create a finished drawing. **(B)**

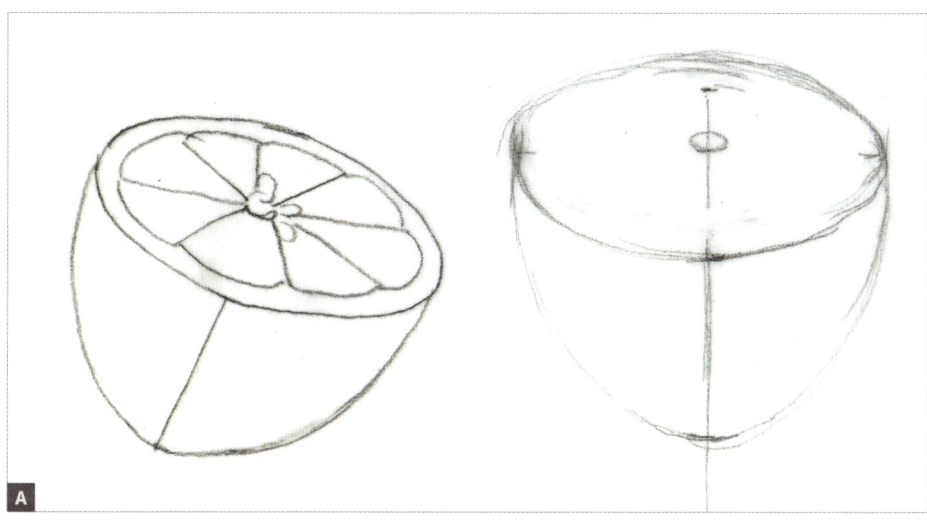

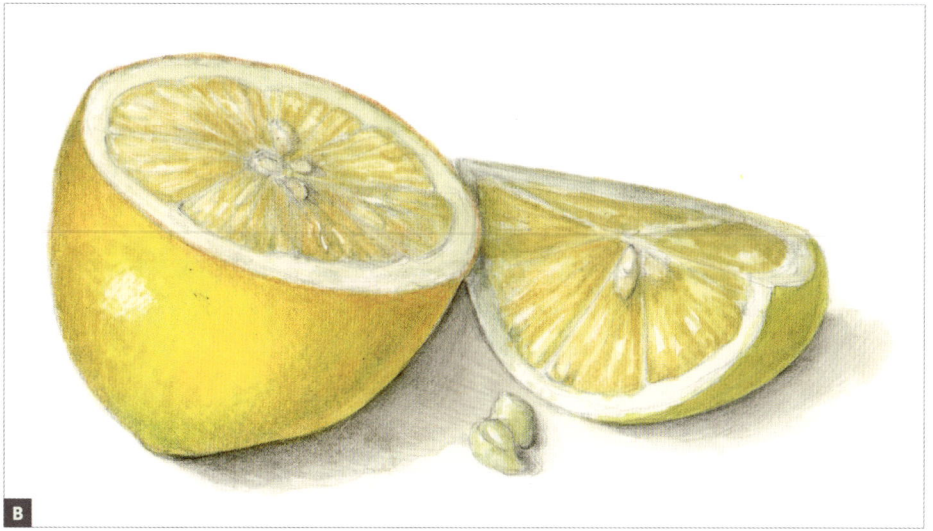

Drawing a Tubular Flower in Perspective

SUBJECT

Tubular flower

ADDITIONAL MATERIALS

- Red Violet colored pencil #194
- Dark Indigo colored pencil #157
- Middle Purple Pink watercolor pencil #125
- Watercolor brush #2
- Colored pencils and watercolor pencils to match the colors of your flower

COLORED PENCILS

#194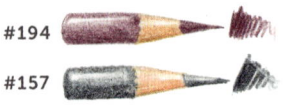

#157

WATERCOLOR PENCILS

#125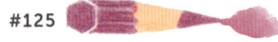

In this lesson, we continue to use ellipses for measuring a flower with proper perspective. I've chosen a pink allamanda, which is a flower shaped like a tube with petals that are joined at the base of the bloom. Use a similar flower if possible. Other subjects that will work fine are daffodils, tulips, campanulas, and lilies.

Look at your flower from various views and do some rough practice sketches to determine the view you want draw. It's good practice to measure and draw three views before choosing one to draw in detail. To keep my flower stable, I often position it in a frog prong. Repeat this lesson as many times as you want with different flower subjects.

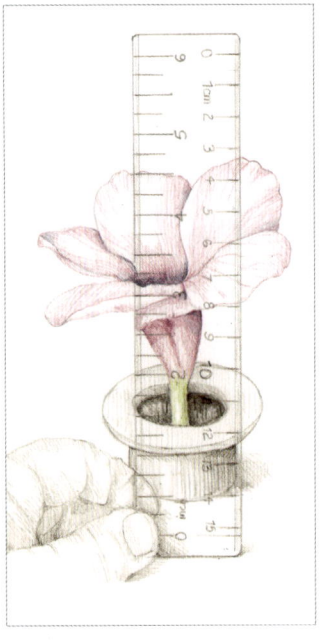
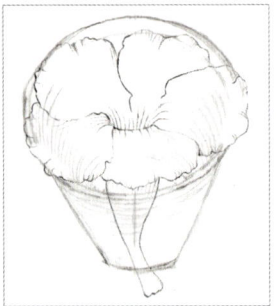
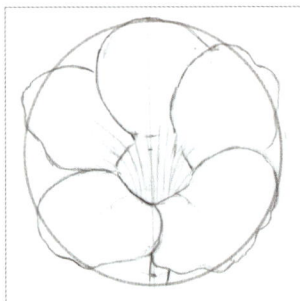
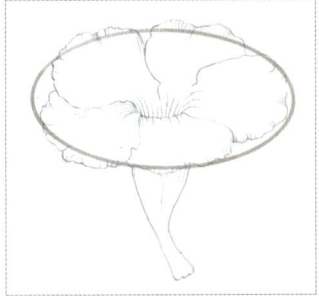

1. Choose a foreshortened view of your flower. Measure it with a ruler on your imaginary window, creating an elliptical view. Lightly draw the flower with graphite pencil. **(A)**

2. Use a neutral shading pencil, such as a Red Violet colored pencil, to add cross-contour lines on the inside of the tube and the petals, describing their three-dimensional bending. Add tone to the overlapping petals and in the deep, trumpet-shaped center of the flower with Dark Indigo colored pencil. Also tone underneath the petals on the outside of the trumpet shape. **(B)**

3. With Middle Purple Pink watercolor pencil and a #2 brush, make a wash of the lightest local color; wet your paper first so that you keep a gradual change in value, but make sure to leave the highlight empty and add a layer of watercolor, leaving the highlights on the petals the color of the paper by carefully adding a wash from the highlighted area toward the shadow. **(C)**

4. Add layers of colored pencil to continue to define the three-dimensional quality and color intensity of the petals. Use cross-contour lines and petal veining by stroking in the direction of these lines to describe the petal veining and surface contour variations. **(D)**

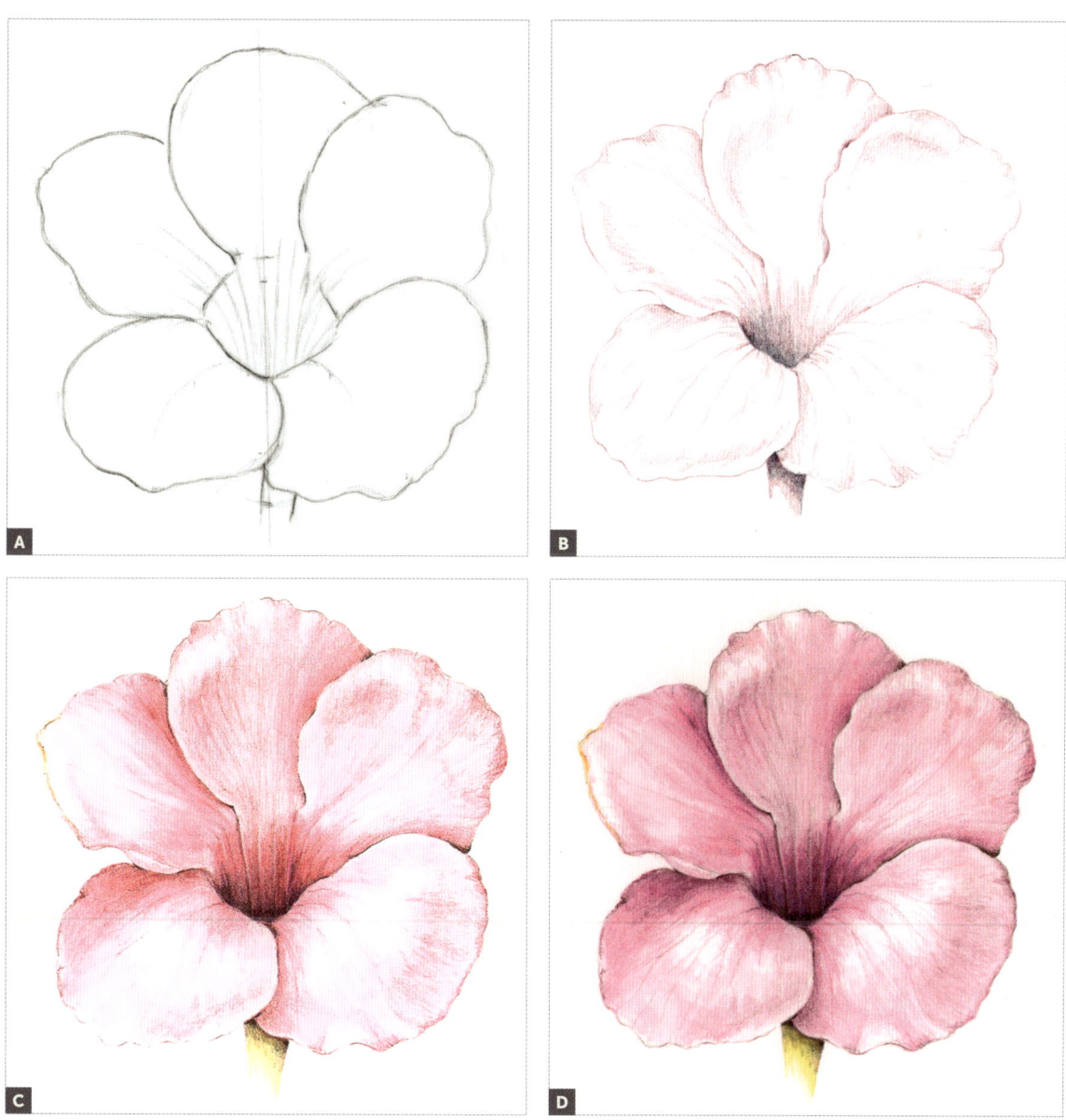

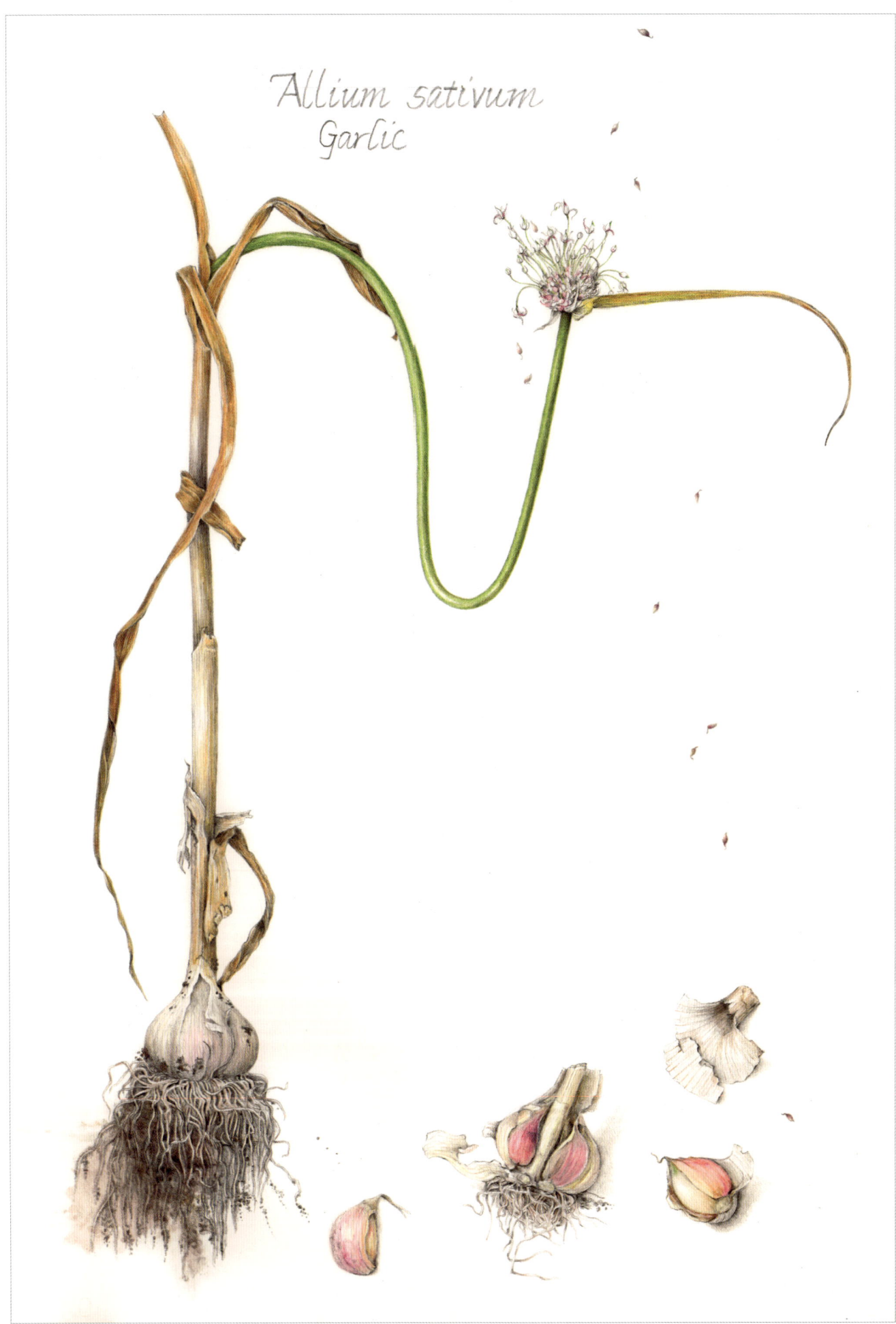

Allium sativum
Garlic

CHAPTER SIX

OVERLAPS

Overlapping is a simple technique that describes complex structures on plants and makes your composition very three-dimensional. An overlap occurs any time one form is in front of or behind another. This can be as simple as one flower petal slightly overlapping another, or as complex as many roots overlapping one another. Drawing through a hidden element and shading behind the form in front will make your overlaps more convincing, as long as all lines seem to continue even when part of a line is hidden. After all, realistic botanical drawing is all about creating the illusion of three-dimensional form on a two-dimensional piece of paper. To enhance this illusion, rendering realistic overlaps is key.

When creating overlaps, especially small ones such as of thorns, hairs, and roots, I always push color out of my mind and think only about what is in front and what is behind. I remember that light elements come forward while dark ones recede, so I make sure that my forms in front are lighter than those behind. Even if the forms in front are dark in color, I push that out of mind. I say to myself, "Well, light could be hitting this form and creating light areas, so therefore I can draw them as light and tone darker behind." Because artists try to describe the structure of a plant, this convention of thinking of light and dark rather than color is used extensively in scientific illustration, much of which is in black-and-white— mostly pen and ink. I figure if it's good enough for scientific illustrators, it's good enough for me!

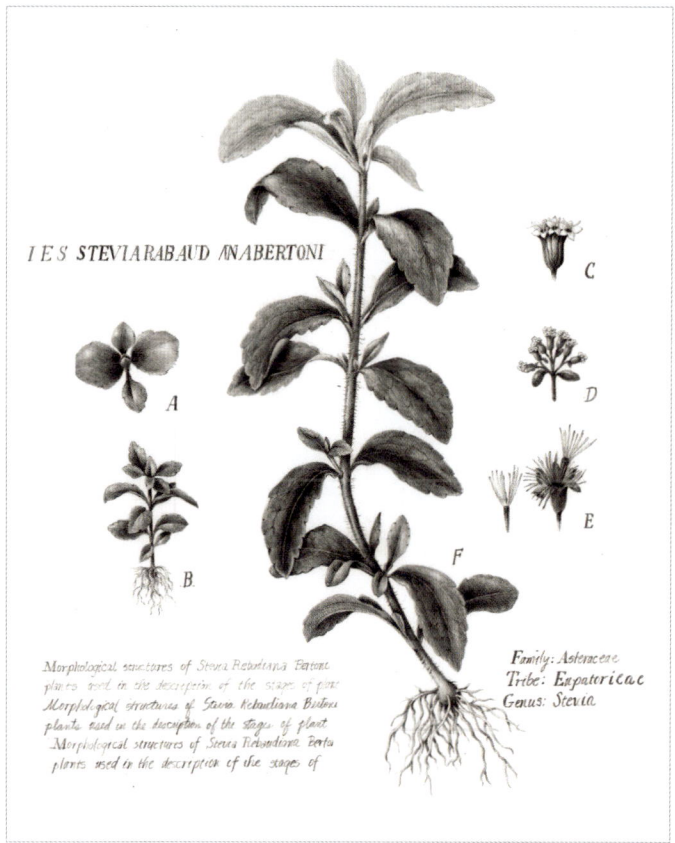

Overlapping Leaf

SUBJECT

Leaf with overlapping edges

ADDITIONAL MATERIALS

- Tracing paper
- Dark Sepia colored pencil #175
- Watercolor and colored pencils to match the colors of your leaf

COLORED PENCILS

#175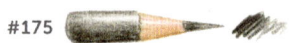

A net-vein leaf that twists and turns is a challenging but exciting subject. The overlapping parts of the leaf create an opportunity for good contrast in light and shadow. If you have access to a curling dried leaf, it will be easier to study how the edges of the leaf overlap. Again, a *Magnolia grandiflora* leaf is perfect for this lesson, because it holds its curled shape indefinitely, but other leaves will work as well.

Here's the secret to a good overlapping leaf drawing: the lines should all connect, even as they disappear.

1. Begin by drawing your leaf lightly with a graphite pencil on tracing paper. Draw the entire center midvein, even if it's hidden from view, and show how it curves. I've drawn all three lines with different colors to keep track of edges that are hidden, indicating the midvein with a red pencil and using a dotted line where it's hidden from view.

2. Draw the edge of the leaf closest to your view. Try to mimic the curve of this edge exactly so that it appears graceful. I've indicated the edge closest to me in green.

3. Draw the back edge of the leaf in one continuous line—even when it's hidden from view. I've drawn this edge in blue. **(A)**

4. Use your tracing-paper sketch as a guide to redraw your leaf on good paper with a graphite pencil. Make any necessary adjustments to ensure that the hidden edges and midvein appear visually connected. Indicate secondary veining. **(B)**

5. With a Dark Sepia colored pencil, tone the overlapping edges first so that there are shadow tones on the areas underneath the fold closest to you. Remember that dark tones recede and light tones visually come forward; this helps create the illusion of three-dimensional form. Tone the bending edges of the leaf to indicate the subtle shadows created when the form bends away from the light source. **(C)**

6. Add a layer of watercolor in the appropriate colors for each side of the leaf. **(D)**

7. Finish rendering the leaf with colored pencils, toning the petiole (stem) like a cylinder, and add in subtle veining and a cast shadow. **(E)**

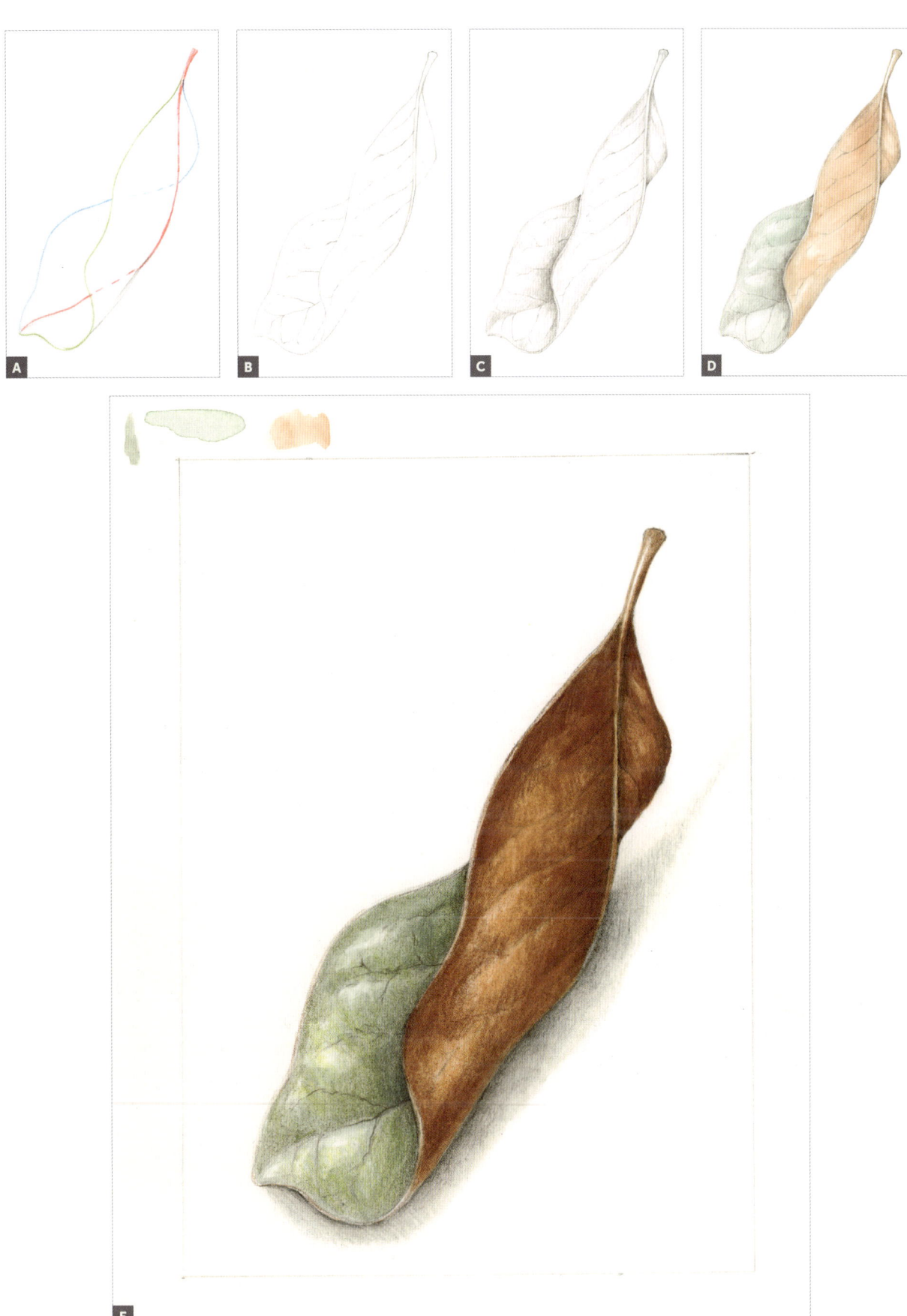

Basic Root Overlaps

SUBJECT

Bulb with roots

ADDITIONAL MATERIALS

- Dark Sepia colored pencil #175
- Magnifying glass or other magnifier

COLORED PENCILS

#175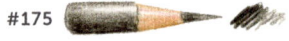

Look closely at a subject that has roots, such as a bulb, to understand the pattern that roots follow. Roots can be irregular, wrinkled, or hairy. Often a root is wider at the base and gets narrower toward the tip, just like veins on leaves. Repeat this lesson with similar subjects, such as other root vegetables or any plant with roots.

1. With a graphite pencil, draw an approximation of a single root: its shape, size, and width variations. **(A)**

2. With a very sharp Dark Sepia colored pencil, carefully redraw the root with a sensitive line (see below), not a solid outline. Look through a magnifying glass to see the variations in your root. Draw a root behind the first root to create an overlap. **(B)**

3. Add tone at the edges where the roots meet to accentuate the idea that these forms are overlapping each other. Add the toning on the root underneath, creating a slight cast shadow. Tone where the roots connect to the bulb. **(C)**

4. Continue to add more roots, one behind the next, and tone them all, showing the overlaps. Remember your light-source model of a cylinder to tone each individual root. Reserve your darkest dark tones for the roots in the back and your lightest lights for the roots in front. **(D)**

SENSITIVE LINE PRACTICE

Sensitive lines is a phrase I use to describe variations in outlines. It's useful to draw sensitive lines whenever depicting thin, undulating roots and edges on leaves and petals. We want to avoid drawing petals that look like solid shapes with hard edges; they should look more like delicate petals with variation in their edges and folds. To accomplish this, vary the pressure you put on your pencil to create lines that feel as if they move around a form, receiving light in places differently, rather than solid, hard outlines. By varying pressure, you can also create wider lines that taper and narrow. It helps to avoid making repeating patterns on your edges; vary your strokes so that things look irregular. Adding a subtle toning inside a petal edge next to a line helps as well.

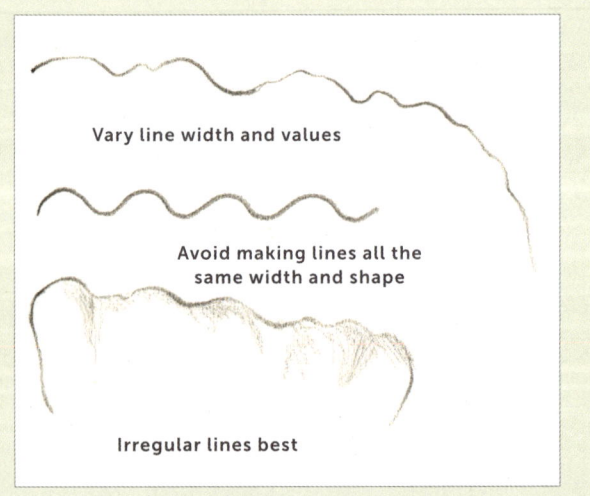

Vary line width and values

Avoid making lines all the same width and shape

Irregular lines best

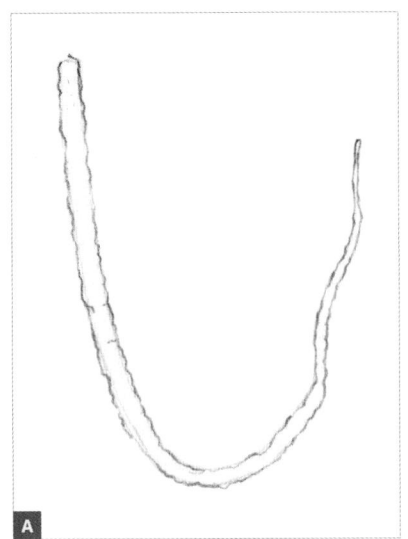

A

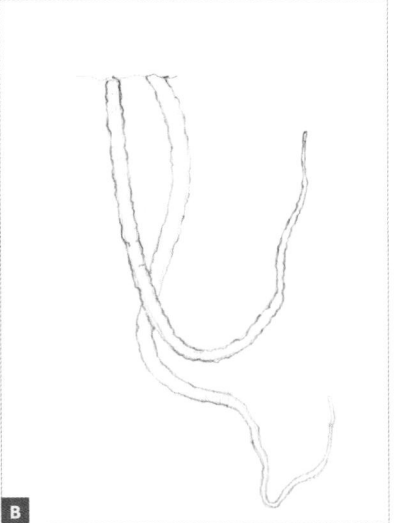

B

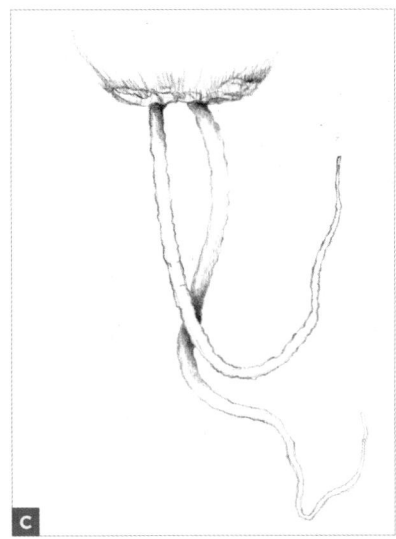

C

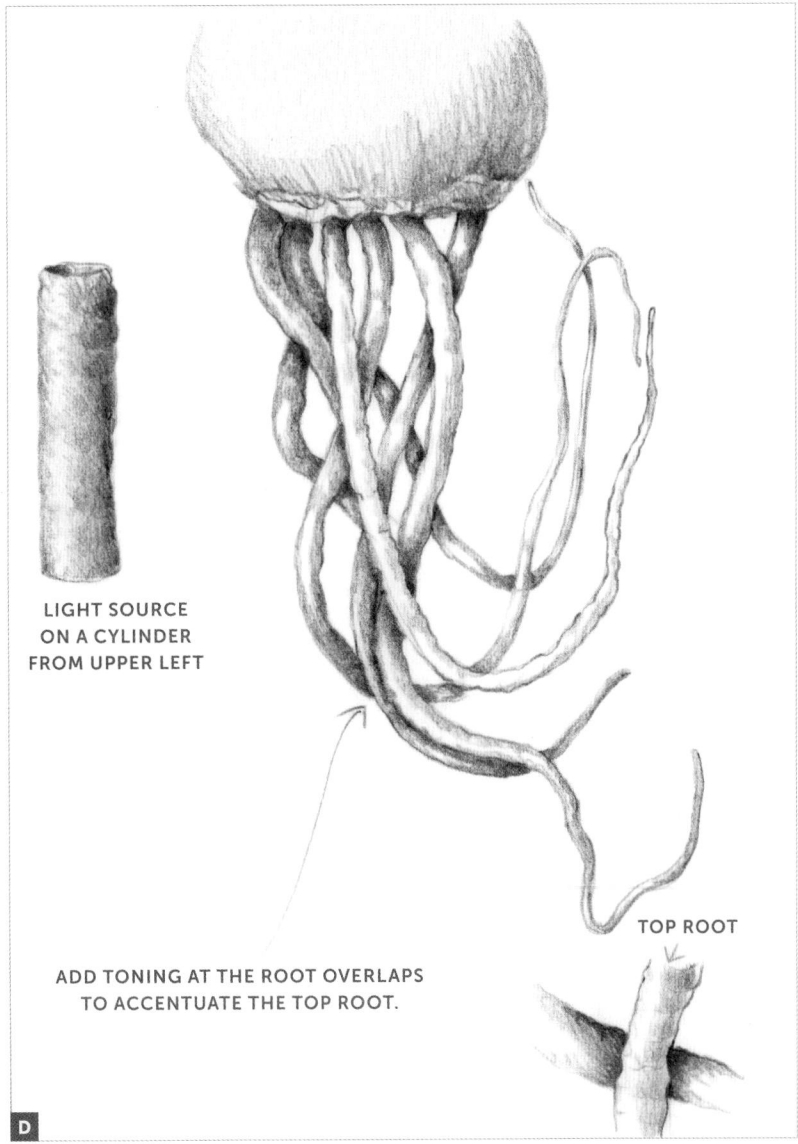

LIGHT SOURCE
ON A CYLINDER
FROM UPPER LEFT

ADD TONING AT THE ROOT OVERLAPS
TO ACCENTUATE THE TOP ROOT.

TOP ROOT

D

Drawing a Garlic Bulb with Thin, Hairy Roots

SUBJECT

Garlic bulb with small hairy roots

ADDITIONAL MATERIALS

- Embossing tools (small to medium)
- Dark Sepia colored pencil #175
- Brown Verithin pencil
- Watercolor brush #2
- Watercolor and colored pencils to match the colors of your bulb
- Wax paper (optional)

COLORED PENCILS

#175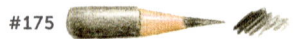

This lesson explores the techniques of embossing on skinny roots. In addition to an embossing tool, you'll use wax paper to lift some of the pencil, creating more subtle roots in the back.

1. Look closely at the roots of your subject and choose embossing tools in appropriate widths.

2. Draw the bulb and roots lightly, leaving plenty of room on the paper for the roots. Start with the roots sitting on top, closest to the viewer, and draw them with an embossing tool. With a Dark Sepia colored pencil, draw over the roots, and add more roots with the embossing tools. Draw some of the thin roots with the pencil as well, not just the embossers. **(A)**

3. Add some more embossing over the darker pencil areas, and again draw over those. You should have a few variations of roots that appear in front and some behind. You can keep building layers of roots this way, creating lots of convincing overlaps and lots of tangled roots. **(B)**

4. Keep building the roots, and refer back to your subject so that the ends of the roots terminate in a natural way. You can draw some roots or the ends of the roots with a Brown Verithin pencil to create variety, some embossed and some drawn. **(C)**

5. Work on the bulb as well, and pay close attention to the area where the bulb meets the roots; render this area nice and dark for good contrast, to create a feeling of mystery, and perhaps indicate some soil clinging to the roots. With a #2 brush, use layers of watercolor—as well as colored pencil—to draw the bulb. Think of the neck as a cylinder and the body as a round form for light source. Remember to have a good highlight on the bulb. Add in veining details that follow the cross-contour lines of the bulb. Create more subtle roots behind with an embossing tool; either do this directly on top of the paper or use the wax paper technique described in step 6 on page 76. **(D)**

6. Use watercolor and colored pencils to darken roots where needed, and to create some good overlapping areas by darkening behind the roots you want to appear on top. Add toning and details on the bulb. **(E)**

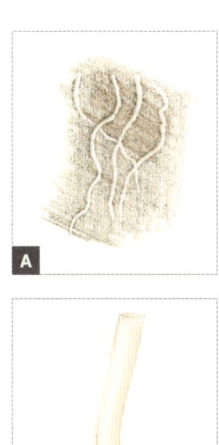

A

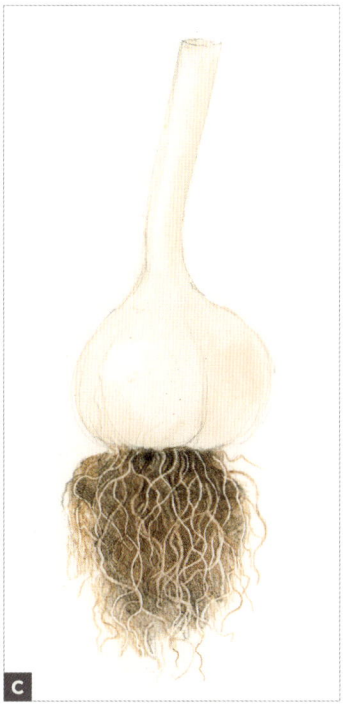

B

C

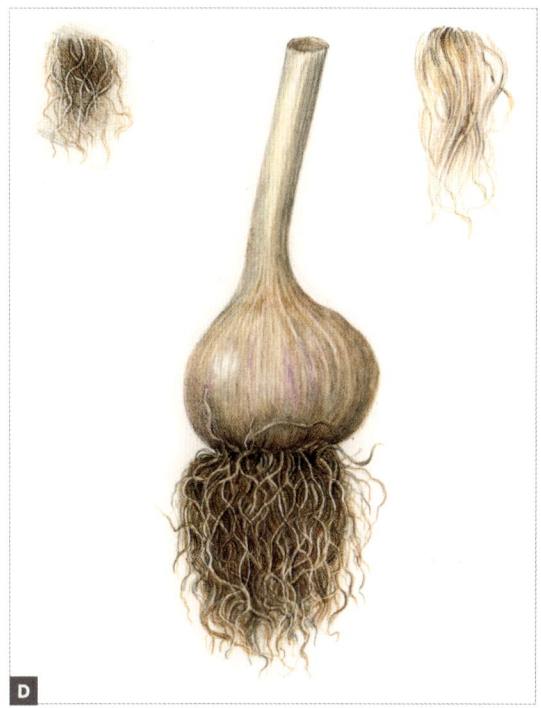

D

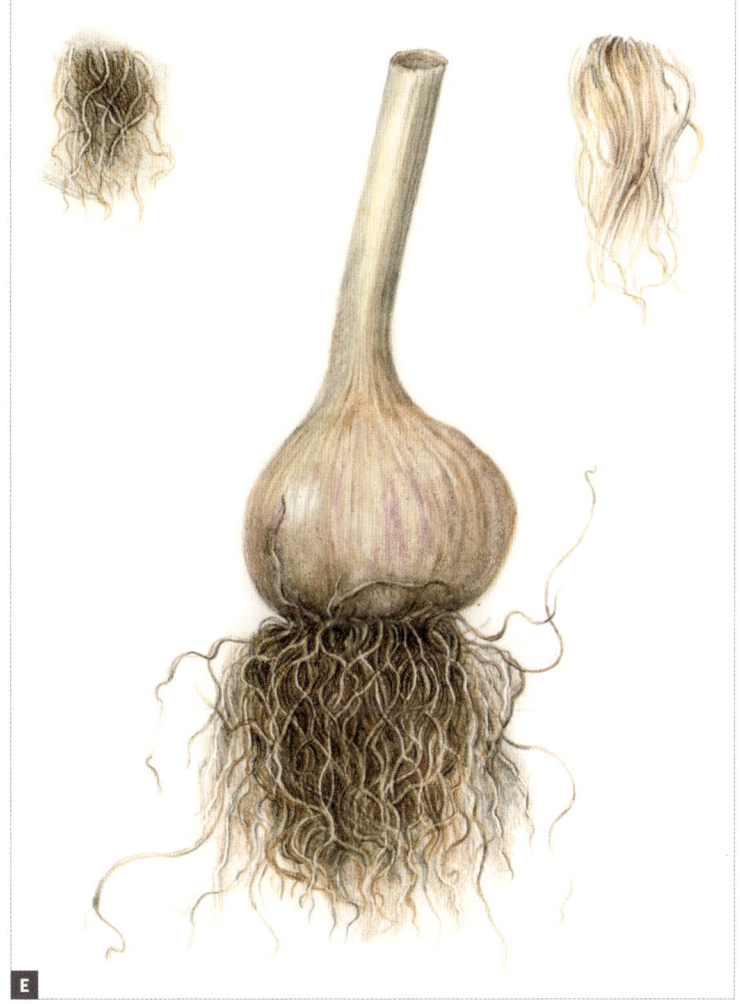

E

Advanced Root Overlaps

SUBJECT

Bulb with overlapping areas of roots

ADDITIONAL MATERIALS

- Dark Sepia colored pencil #175
- Embossing tool (small)
- Burnt Sienna colored pencil #283
- Other colored pencils and watercolor pencils to match the local color of your bulb
- Burnt Sienna watercolor pencil #283
- Watercolor brush #2 or #6
- Permanent Green Olive watercolor pencil #167
- Brown Verithin pencil

COLORED PENCILS

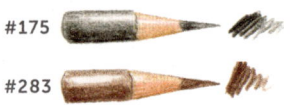

#175

#283

WATERCOLOR PENCILS

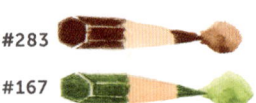

#283

#167

I recently received an amaryllis bulb as a gift. It arrived ready to plant and grow an amaryllis flower, but before I planted it in a pot, I decided to draw it. This is a great way to practice drawing overlapping roots and a round form at the same time. (To see the blooms from this plant, turn to page 177.)

Study a root and notice that it is not straight or a smooth curve, but perhaps a bit bumpy on its edges. A root is usually wider at the base and tapers and gets thinner and thinner at its end. The key to drawing realistic roots is capturing their irregular quality. Try not to approximate this idea, but study and draw it realistically. I allow for variation in my line thickness and values by varying the pressure on my pencil. Practicing line variation is quite relaxing. Your goal in this lesson is to give the roots many layers of depth and to make your bulb in the background very three-dimensional. You don't need to follow your model exactly once you've created the correct pattern of the roots in terms of width variation, edges, and values as they bend away from and toward light.

1. With a graphite pencil, draw a life-size, light outline of the bulb and overlapping roots. No need to draw every root, just an idea of where they will go with an overall composition. **(A)**

2. Switch to a Dark Sepia colored pencil and carefully draw the roots that will appear in front. Continue to add in more roots. Consider allowing some roots to overlap others, and allow the lines to disappear at the overlaps.

3. Now use an embossing tool to add some skinny, hairy roots. Remember to draw sensitively with the embosser so that the lines look graceful and also have variation. Then use the side of the Dark Sepia to lightly tone a bit so that the embossed roots start to appear. You can tone over the embossed roots, but draw around the thicker roots in front so that they remain lighter in value. Begin toning the bulb, remembering to vary the tone values as on a round form with a clear light source coming from the upper left. **(B)**

4. Add layers of Burnt Sienna colored pencil (or whatever the local color of your bulb is) over the Dark Sepia. With small strokes and a sharp pencil, build the layers slowly, pressing softly in the beginning.

5. With Burnt Sienna watercolor and a #2 or #6 brush, add a layer of watercolor on the bulb to start to push the roots to the front. With Permanent Green Olive watercolor, add a watercolor wash to the developing stems at the top of your subject. **(C)**

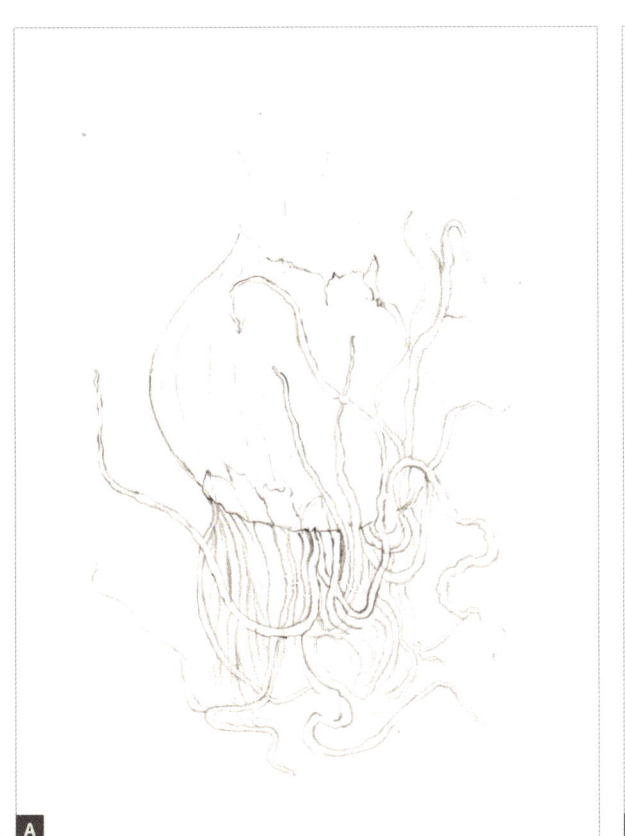

A

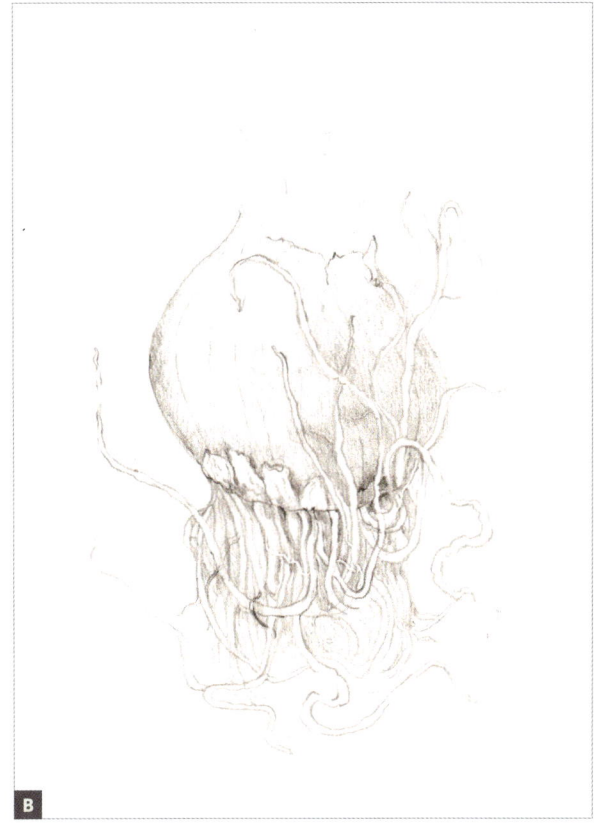

B

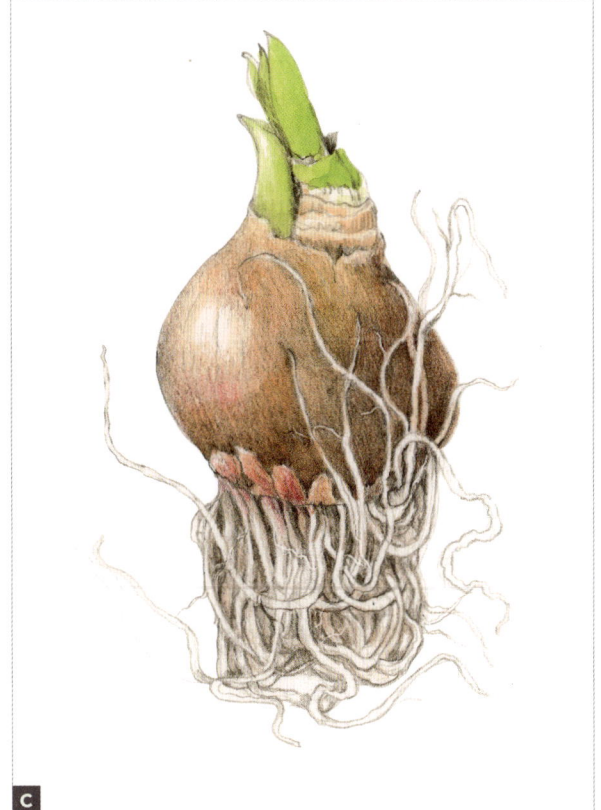

C

CONTINUED

6. Add more toning on the roots using a Brown Verithin pencil, which provides a soft color with sharp control in detailed overlapping areas. Continue to darken and add more tone on the areas behind those in front, whether on a root or the bulb. If there are other colors in your bulb or root, begin to indicate those. I've shown the veining visible on my bulb skin.

7. At this stage, you can create more depth and many overlapping layers of roots by varying the tone values on the roots, darkening those roots behind the roots you want to appear on top. This is the time to build your layers slowly and have fun with artistic license. No need to copy exactly. **(D)**

8. Continue to layer in more value variation to emphasize the overlaps, creating depth and the feeling of a tangle of roots. Work on the bulb surface behind the roots to help give dimension to the roots lying on it. **(E)**

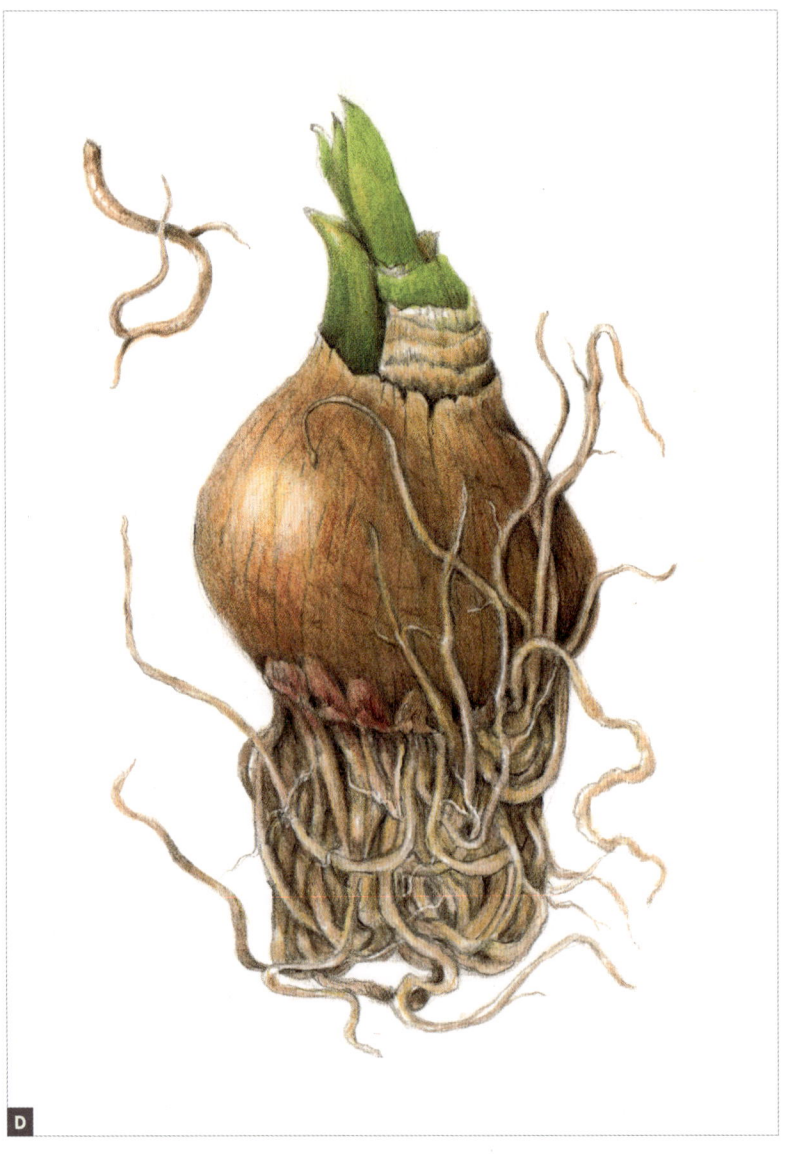

D

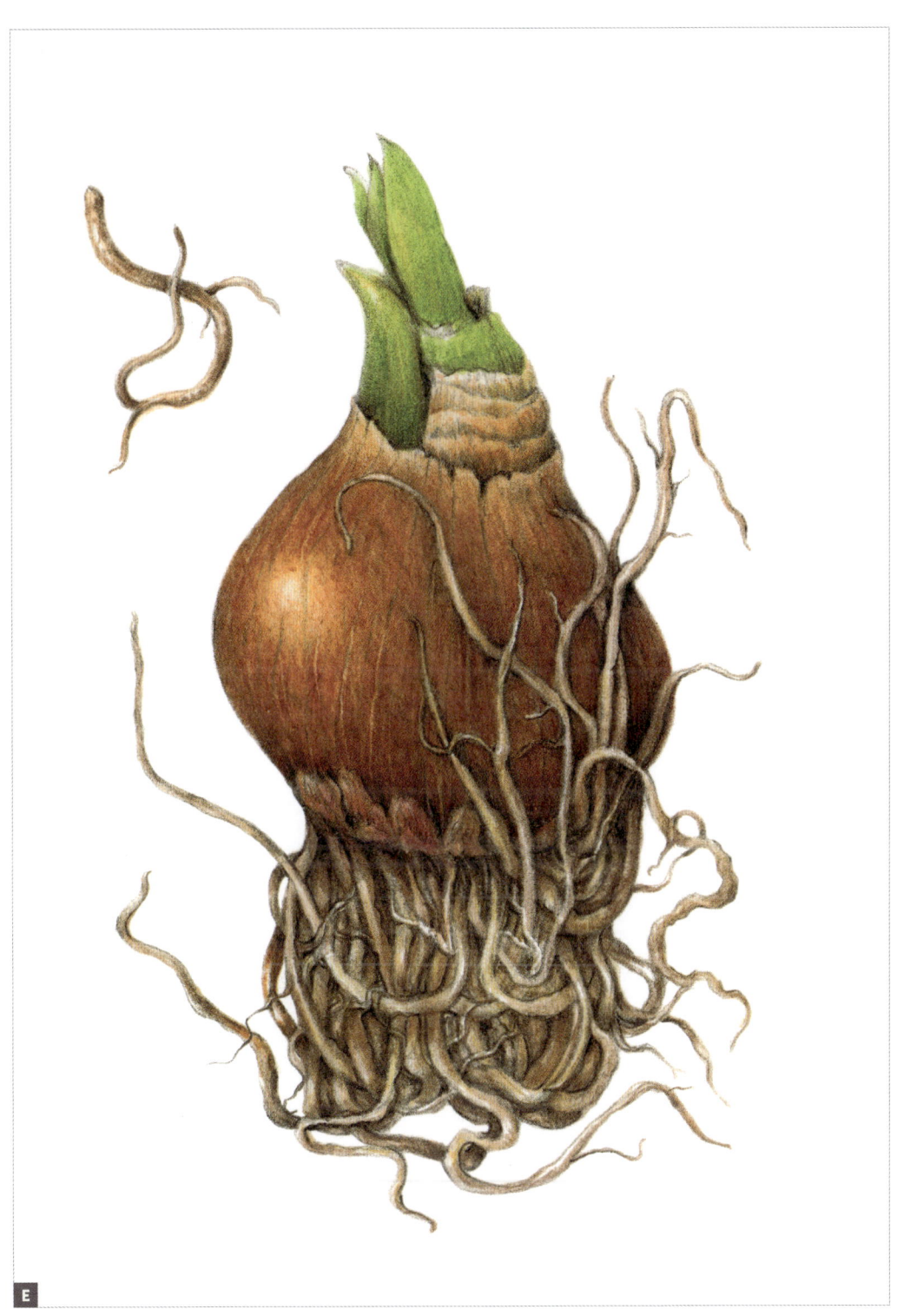

Overlapping Petals of a Flower

SUBJECT

Curvy or rolled flower petal

ADDITIONAL MATERIALS

- Tracing paper
- Colored pencils and watercolor pencils to match the local color of your petal
- Watercolor brush #2
- Dark Sepia colored pencil #175

COLORED PENCILS

#175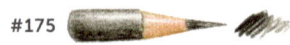

Another way to practice overlaps that are similar in form to a leaf that curves over on itself is to practice the same drawing technique on flower petals. I've found good overlaps on a *Hibiscus elatus*, which has naturally curvy petals (see the complete flower on page 179). This lesson uses the same toning technique for rolling and overlapping forms that we practiced previously with a leaf (see page 94).

1. Use tracing paper to practice drawing the three important overlapping areas: the two sides of the petal and the center midvein (even if the petal doesn't have a prominent midvein). It's helpful to draw this with three different colors to keep track of the sides. **(A)**

2. Redraw your petal on good paper, with a pleasing, convincing curve to your overlaps. **(B)**

3. Begin to add tone with a neutral colored pencil (in my case, Red Violet) behind the areas of the petal that are on top. Define a good highlight on the top. If your petal has strong veining that follows the surface contour, be sure to stroke in this direction, as it will enhance the illusion of the petal overlapping and continuing on the underside. **(C)**

4. Using a watercolor brush #2, mix up a local color of watercolor, and add a layer of watercolor wash leaving a good highlight. **(D)**

5. Finish your drawing by continuing to add layers of colored pencil that match the local color of your subject. Emphasize details you see on your subject with toning and different colors. Use Dark Sepia colored pencil for dark shadows and burnish to blend the layers together, leaving a good highlight where appropriate. **(E)**

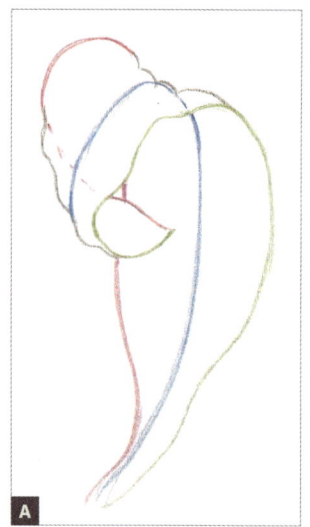

A

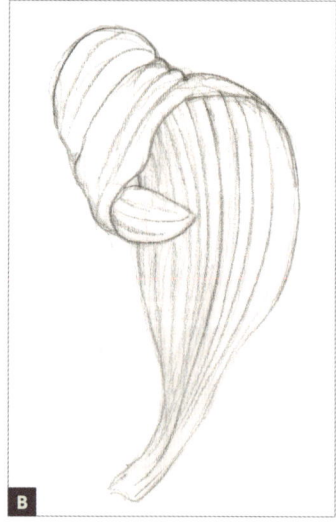

B

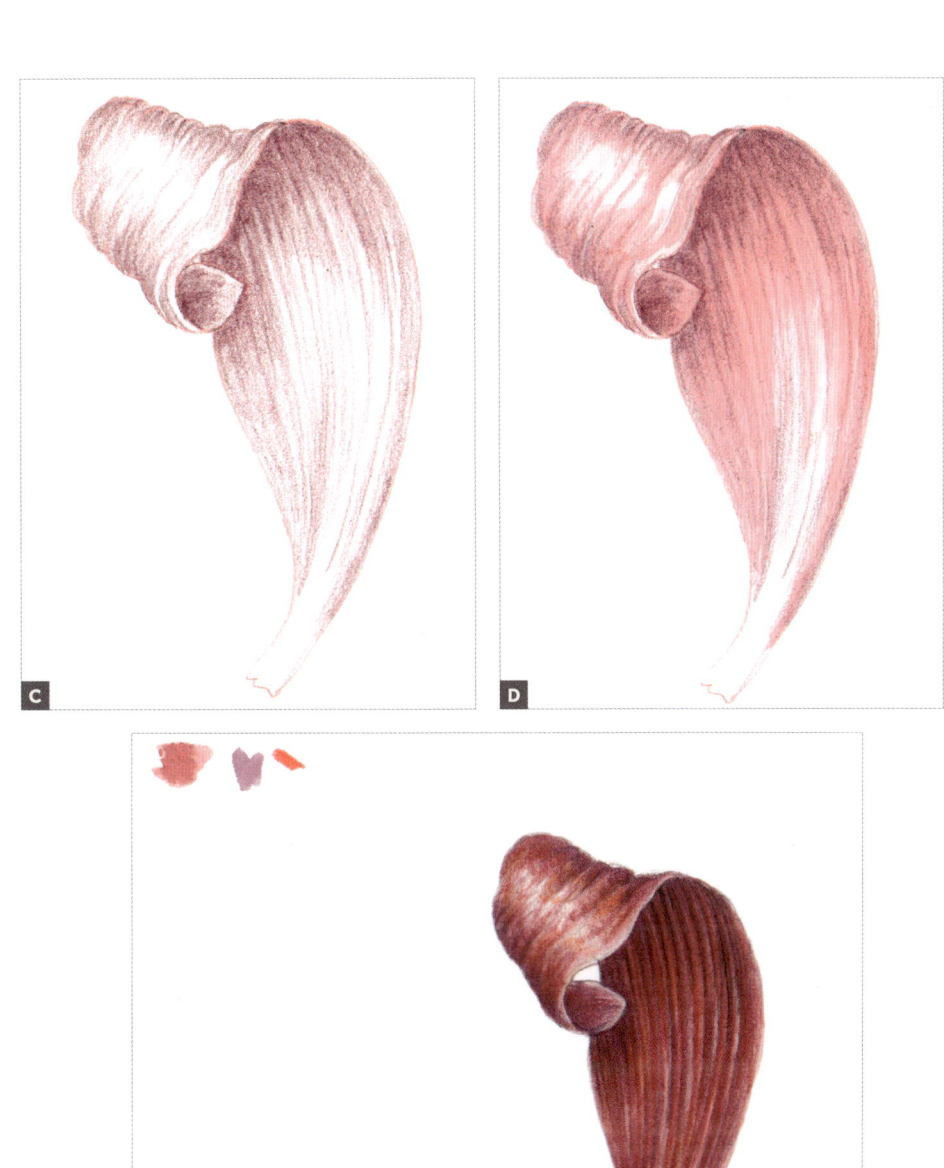

C

D

E

Overlapping on a Rolled Piece of Birch Bark

SUBJECT

Rolled or curled piece of bark

ADDITIONAL MATERIALS

- Tracing paper
- Colored pencils and watercolor pencils to match your bark's local color
- Dark Sepia colored pencil #175
- Watercolor brush #2

COLORED PENCILS

#175

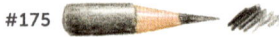

Once you've created several kinds of overlaps, it's fun to practice on a piece of curling bark. Curling bark contains a spiral pattern common in many forms. I am fondly reminded of Leonardo da Vinci's drawings of a small child with curly hair. Perhaps I'm so fond of these spiral patterns because I have extremely curly hair, and it's fun to study and draw the corkscrew pattern. Tendrils on plants with vines also display the same spiral patterning, and often they conform to a cylinder shape.

1. Find an interesting piece of rolled bark for this lesson. Use a light-source model of a cylinder, and draw a thumbnail sketch as a reminder of the surface contour and the light source you'll use.

2. Draw your rolled bark on tracing paper. Follow the edge closest to you, and draw with a colored pencil to trace the edge as it curls around. **(A)**

3. With a Dark Sepia colored pencil, draw your entire subject on good paper and shade the overlaps. Shade the areas underneath the top of your subject. Good contrast is important, and you can create contrast by juxtaposing very dark areas with light areas on top of them. **(B)**

4. Using a watercolor pencil that matches your subject's local color and a #2 brush, add in layers of watercolor washes to indicate color variations, highlights, and shading. **(C)**

5. Finish your drawing by continuing to add layers of local colored pencil. Emphasize details you see on your subject with toning and different colors. Burnish to blend the layers together, leaving a good highlight where appropriate. Add more layers to distinguish details and add a cast shadow if you'd like. **(D)**

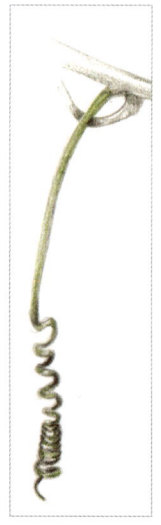
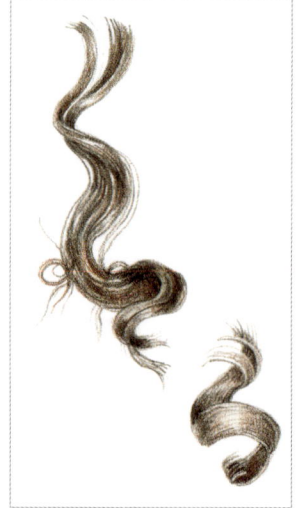

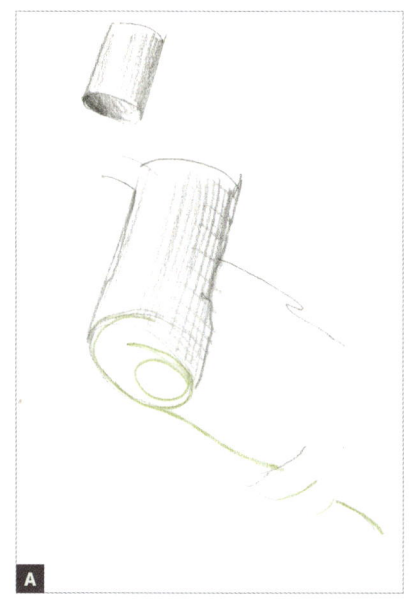

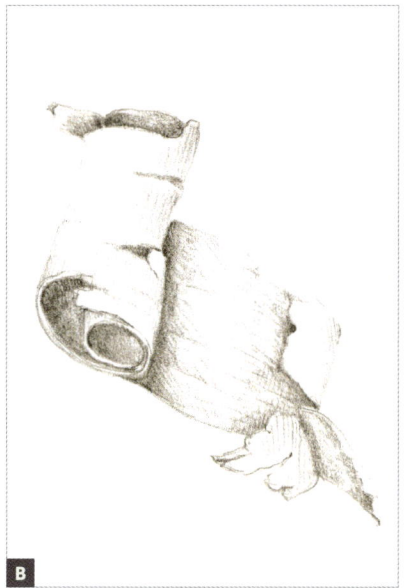

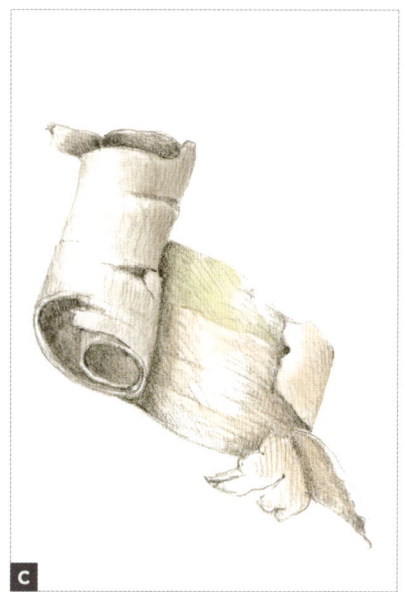

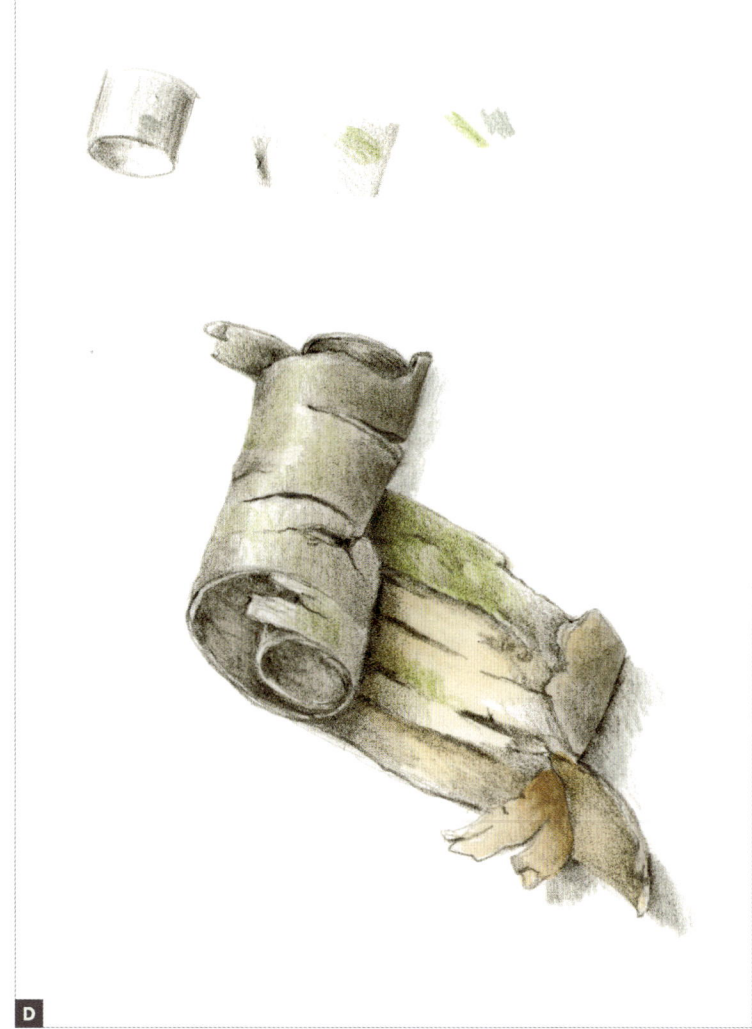

Overlapping on Small Fruits, Leaves, and Stems

SUBJECT

Any small fruits with stem and leaves attached

ADDITIONAL MATERIALS

- Tracing paper
- Dark Sepia colored pencil #175
- Red Violet colored pencil #194
- Colored pencils and watercolor pencils to match the local color of your subject
- Watercolor brush #2

COLORED PENCILS

#175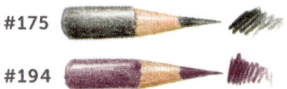

#194

In this drawing of plums, I used the concept of overlapping several times. Utilizing overlaps is a great way to create contrasting areas, which can give your drawings a good sense of depth. By darkening the area behind a form, that main subject is pushed forward (advancing toward the eye) while the other elements recede, and this creates a focal point—the spot on which you'd like viewers to focus. We've practiced drawing spheres, leaves, and cross sections. Now we're going to put these elements all together in a small composition. Hopefully you're excited to do this, just as I always am! When you start to gain confidence in one area, you will naturally want to keep adding on to your newfound skills.

1. On tracing paper, plan a small drawing with multiple fruits, leaves, and stems to practice overlapping. Indicate with toning where your overlaps will be.

2. With a graphite pencil, draw the subject with all the elements on good paper. **(A)**

3. With neutral colored pencils such as Dark Sepia and Red Violet, tone your overlaps. Make sure to tone the surface contour of the forms realistically where the shadows sit; try to avoid a dark, flat shadow on the fruit by varying your shadow values.

4. Once your overlaps are clearly established, continue your drawing with grisaille toning, remembering to use a clear light source. **(B)**

5. Use a #2 brush to add local color with watercolor layers. **(C)**

6. With colored pencils that match the local color of your subject, add layers of color. Darken and layer the colors as needed to retain good contrast between the overlaps.

7. Optional: Add elements to the composition such as a cross section of the fruit. **(D)**

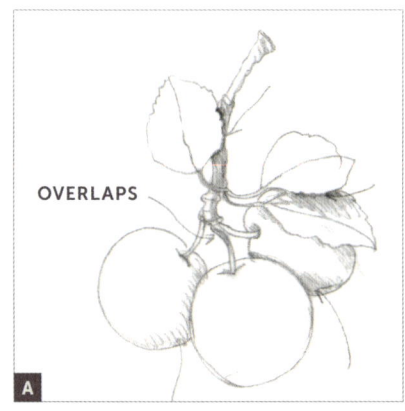

OVERLAPS

A

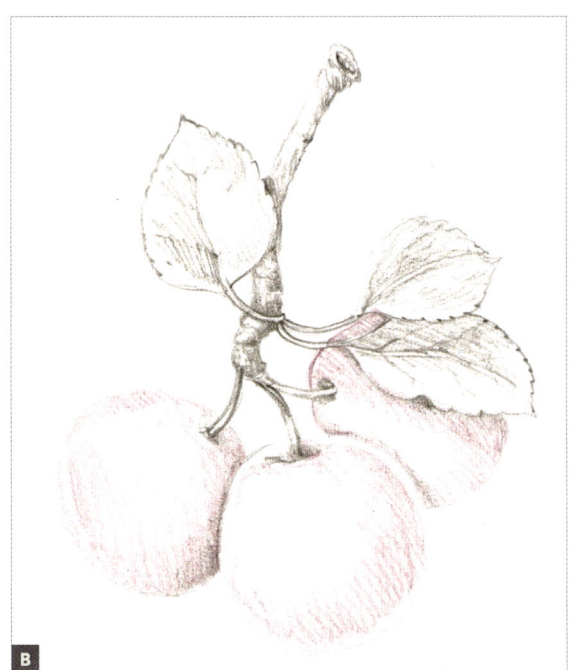

B

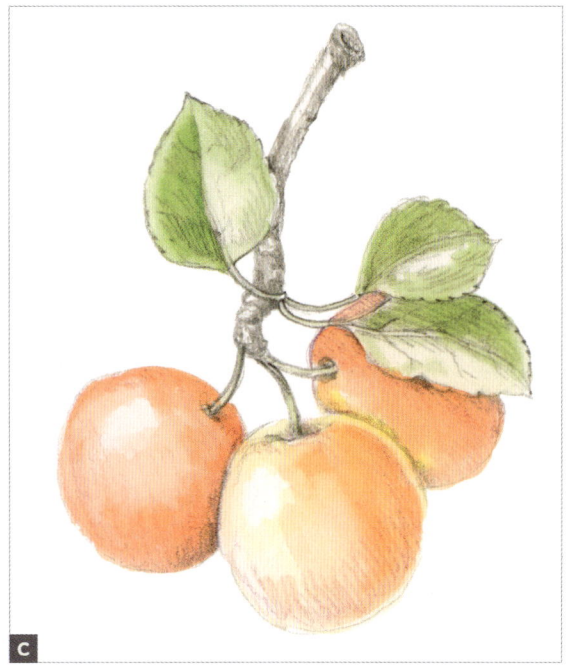

C

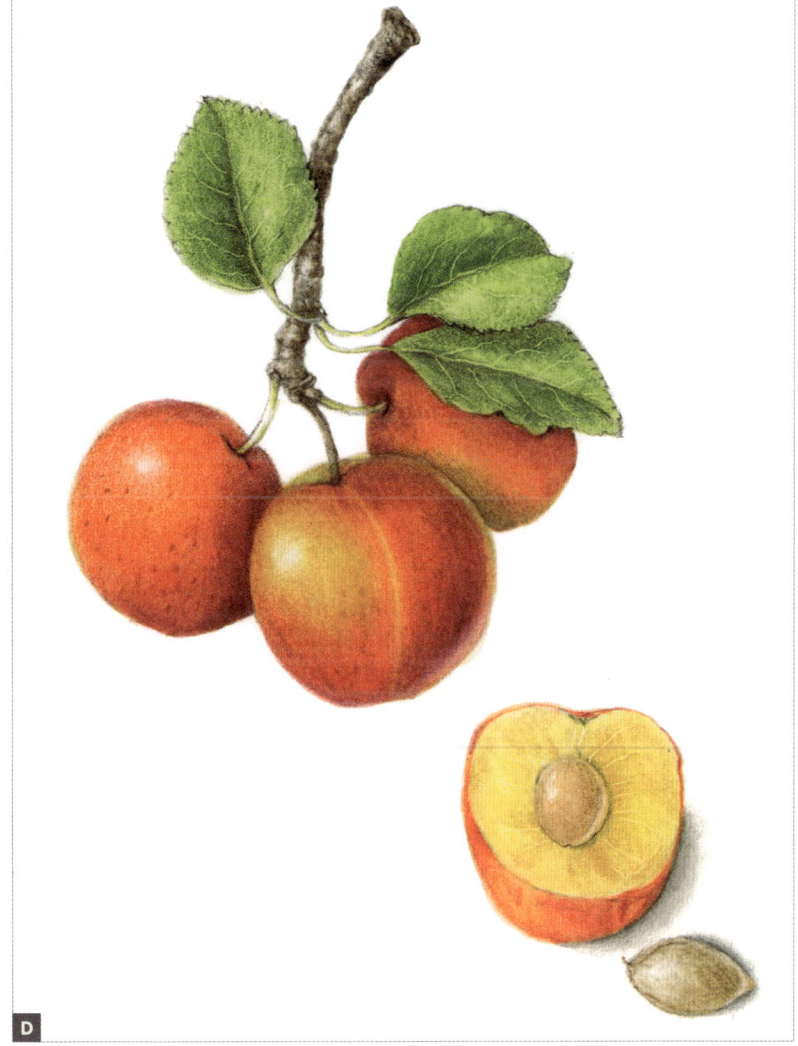

D

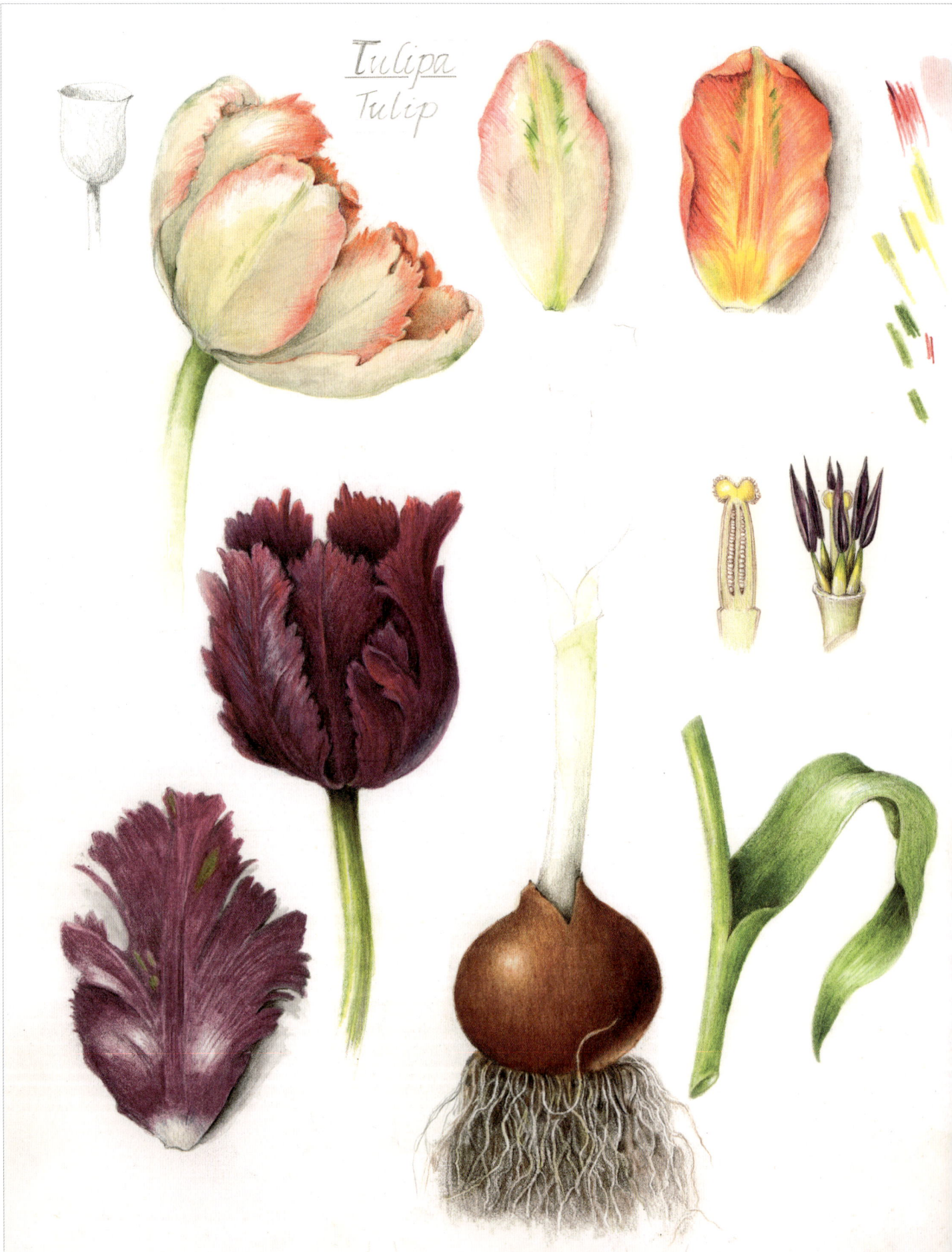

Tulipa
Tulip

CHAPTER SEVEN

EXPLORING FLOWERS

This chapter presents some basics of botany that will support your drawing skills. Take this slow, and imagine you're a botanist as you begin your study of flowers.

Have you ever wondered what a flower is and what its purpose is? I know this may sound silly. We're all familiar with flowers, and we love them for their beauty, but it's important to note that a flower has a purpose beyond looking pretty. But looking pretty is indeed important, because the main function of a flower is attraction. The enticing form, color, smell, and promise of sweet nectar make resistance difficult for any pollinator or person. Pollinators pollinate a flower's ovules, which will then grow into a seed-bearing fruit or seed capsule. Flowers are crucial to the cycle of life. A flower is the reproductive or seed-bearing portion of a plant. It consists of male and female reproductive parts that are sometimes but not always on the same flower. Some flowers have only male or female parts. What follows is a list of basic flower parts; if you're a botany enthusiast, please refer to a book on botanical terms to expand your vocabulary.

SEPALS are leaflike, usually green structures that hold the flower together.

PETALS are often brightly colored and surround the reproductive parts of a flower. If sepals and petals look alike, they are collectively called *tepals*.

The **PISTIL** is the female part of the flower containing the *stigma*, *style*, and *ovary*. Inside the ovary are *ovules*, which are the immature seeds.

The **STAMEN** is the male part that contains a *filament* and an *anther*, which holds the pollen used to fertilize the ovules.

During fertilization, a pollinator inadvertently collects pollen on its body when in search of sweet nectar, which is located in the center of the flower. The pollinator moves from one flower to another. Pollen will land on the stigma, work its way through the style, and fertilize the ovules, which then develop into a fruit containing seeds. The seeds can grow into a new plant and continue the cycle of the plant year after year. I love to compare the structure of an ovary with the fruit or seed

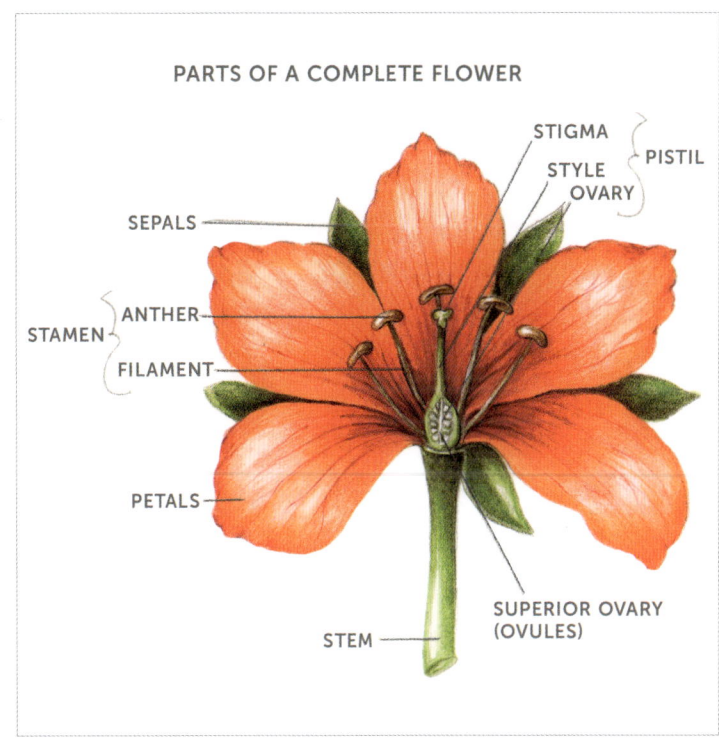

PARTS OF A COMPLETE FLOWER

STIGMA
STYLE
OVARY
PISTIL
SEPALS
ANTHER
STAMEN
FILAMENT
PETALS
SUPERIOR OVARY (OVULES)
STEM

capsule that later develops. I often include these dissections in my drawings.

Once you've learned the basics of drawing several types and shapes of flowers, it's time to explore all the components of a flowering plant. No need to feel you must do finished botanical drawings. The most important thing is that you practice, practice, practice! I love my plant process pages, where I can just take apart flowers, draw, and explore color and whatever else strikes me about my flower.

Understanding and drawing flowers is a lifelong pursuit, and aren't we lucky to get to do this?! In this chapter, we'll be drawing flowers in combination with other parts of the plant over and over again, so there's no hurry to understand it all in this first lesson. Just explore and enjoy.

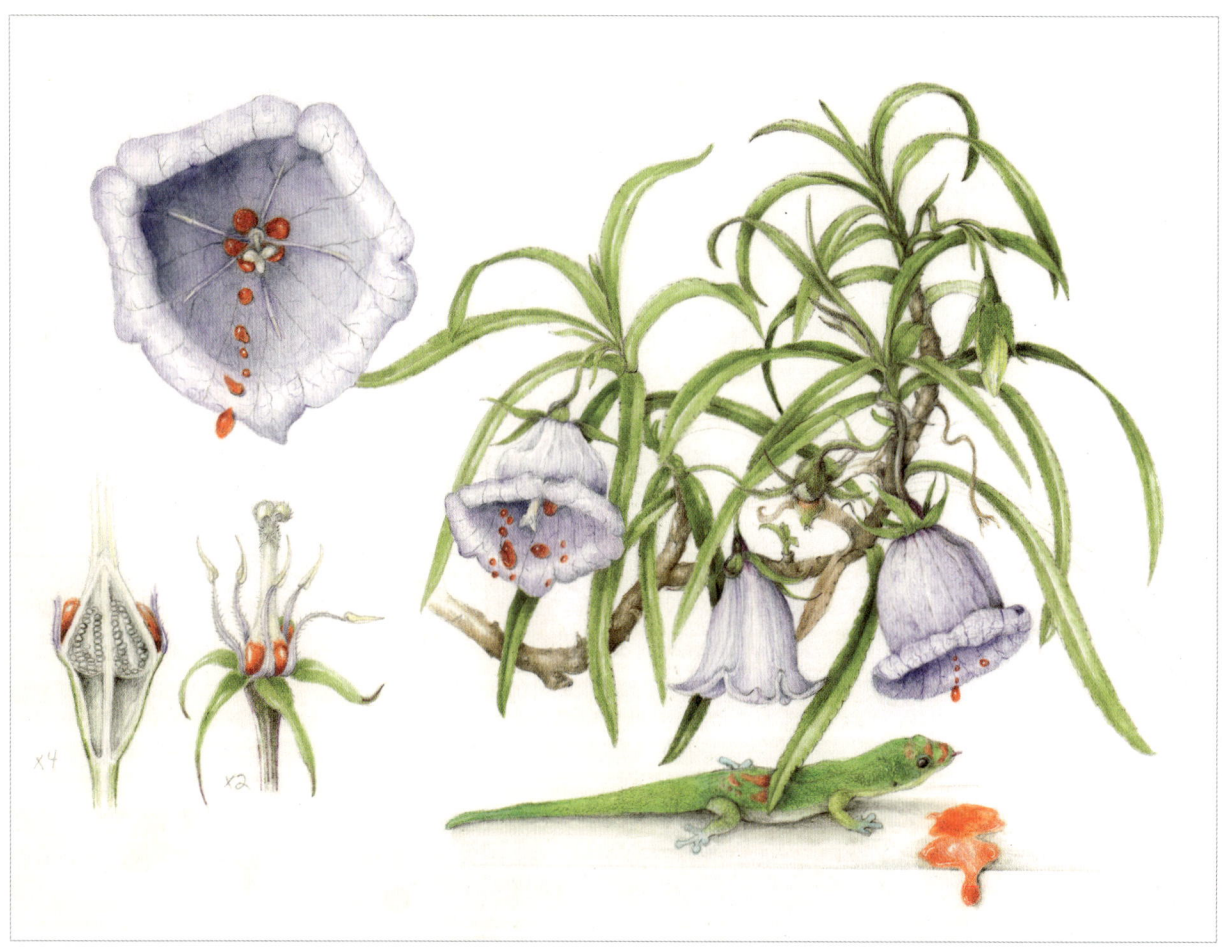

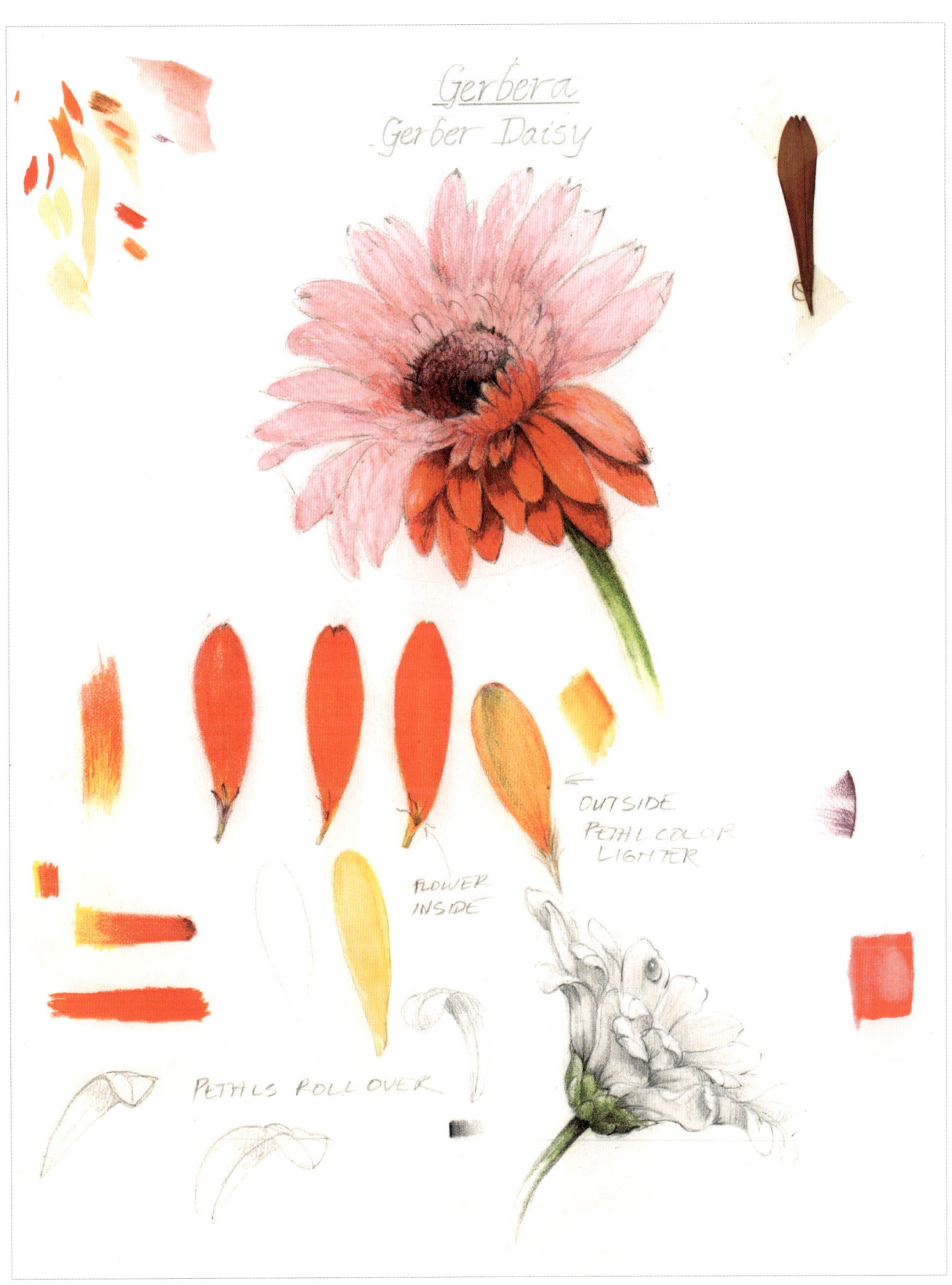

Gerbera
Gerber Daisy

OUTSIDE
PETAL COLOR
LIGHTER

FLOWER
INSIDE

PETALS ROLL OVER

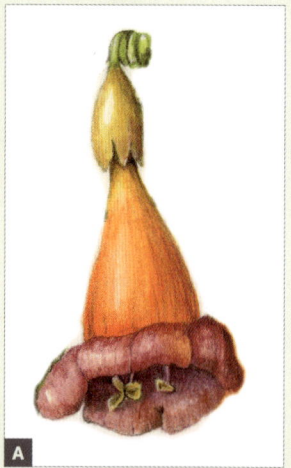

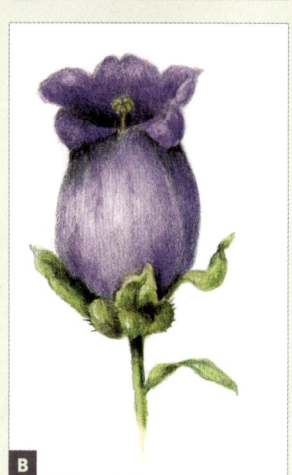

FLOWER SHAPES SIMPLIFIED

Thinking about the basic shapes of flowers and their symmetry is helpful when drawing an entire flower. Most flowers can be simplified into a few basic geometric shapes. Look at a flower and identify which simple shape is most like it. Tubular flowers are one of the simplest flower shapes. The petals are joined to form a tube either entirely or partly, which is a good shape for drawing, because the three-dimensional form is very evident.

TUBULAR OR TRUMPET SHAPE: A flower with a tube formed of united petals, often separating at the mouth into a flared shape where the petals often curl back, such as a trumpet vine and an allamanda. **(A)**

CAMPANULATE OR BELL SHAPE: A flower with a wide tube and flared petal tips, typical of the bellflower family, such as a nectar campanula or other campanula. **(B)**

FUNNEL FORM OR FUNNEL SHAPE: A flower that widens gradually from the base, ending in an open or flared shape, such as lily, morning glory, or azalea. **(C)**

CUP SHAPE: A flower that widens gradually from the base and is formed by individual petals (collectively known as tepals). The tepals conform to a cup shape, which can be seen in flowers such as a tulip or crocus. **(D)**

ROTATE OR ELLIPTICAL SHAPE: A disk-shaped flower that is mostly flat and circular, such as daisies, anemones, and sunflowers. **(E)**

COMBINATION SHAPE: A trumpet shape and a rotate or elliptical shape together, such as a daffodil. **(F)**

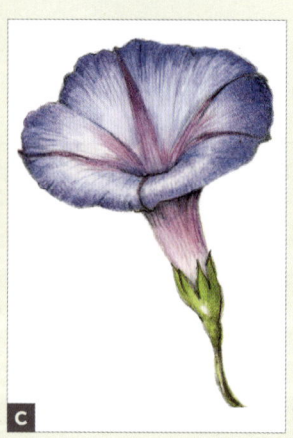

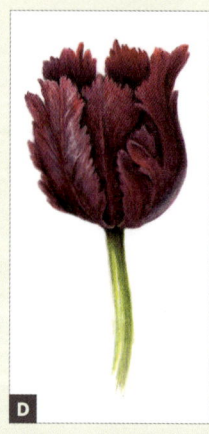

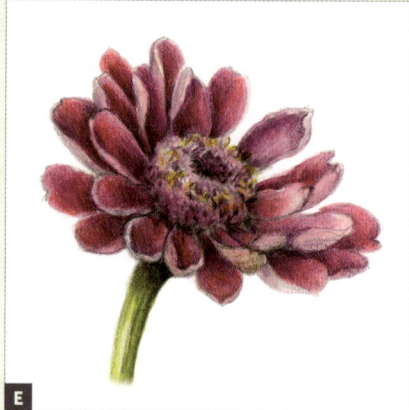

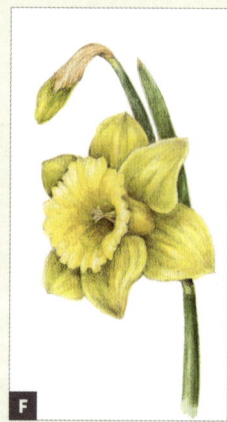

FLOWER SYMMETRY

Many flowers are symmetrical and can be segmented into identical sections. Some blooms have no axis because they grow in a spiral. Knowing the symmetry of a flower makes it easier to draw.

RADIAL SYMMETRY: The arrangement of parts is around a single main axis that will produce a mirror image on the plane at any angle when dissected in half. Think of the spokes on a bicycle wheel. Flowers such as zinnias and anemones have radial symmetry. **(A)**

BILATERAL SYMMETRY: The structure is a mirror image on either side of a line drawn vertically through the middle. Flowers that have bilateral symmetry include wisteria, orchid, mouthy flowers in the mint family, lavender, and wild bergamot. **(B)**

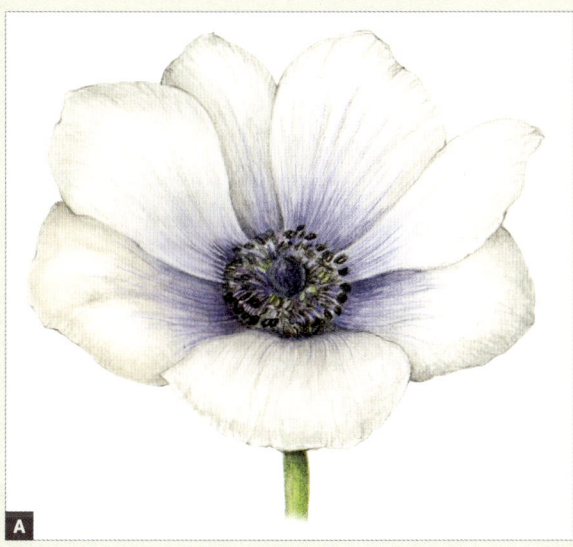

A

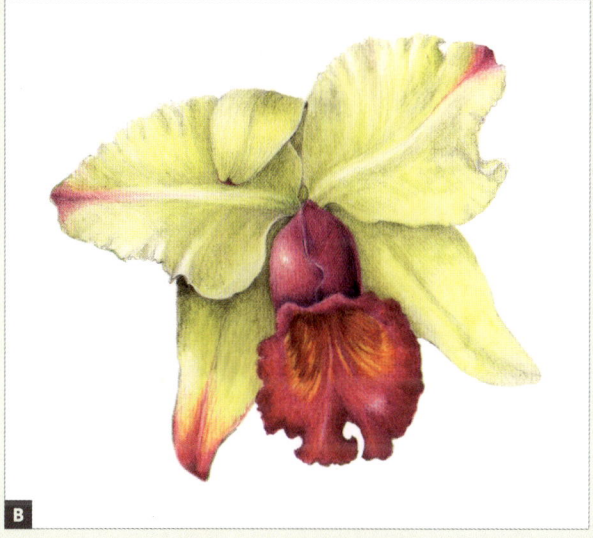

B

Drawing a Petal

SUBJECT

Any petal, about 2 inches high, from any large flower, such as a tulip, daffodil, lily, rose, or iris

ADDITIONAL MATERIALS

- Watercolor brush #2
- Watercolor pencils and colored pencils to match the local color of your petal
- Gray Verithin pencil
- Black Verithin pencil
- Ivory colored pencil #103
- Dark Sepia colored pencil #175

COLORED PENCILS

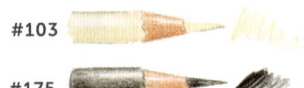

#103

#175

Let's start simple with a single petal from a flower. (The tulip petal I've used in my drawings is a wonderful choice.) This is a good way to study a flower, enjoy drawing and color matching, and closely inspect nature's details. You'll also practice the colors, patterns, and shadows you might use when drawing a whole flower. Another reason to do this is because it's just so much fun. I could do a petal a day endlessly and be happy, even if I never draw that particular flower in its entirety. Drawing a petal works as a daily meditation. If you have a busy schedule, consider making a petal a day your drawing practice, at least in the spring and summer. (See page iv for a complete drawing of the tulip.)

1. Carefully choose a petal that isn't too small and pull it off of your flower. Place the petal on your page, and with a graphite pencil, lightly draw it life-size. Set up a good light source, and note where the highlights appear.

2. Use a #2 brush to paint a watercolor wash as your first layer, leaving the highlights blank. **(A)**

3. If your petal has variegated color, lightly draw in the color variation with colored pencils, feathering one color into the next to keep it subtle. Look closely at your petal's variegation and try to copy it. As you draw, follow the veining pattern direction, which mimics cross-contour lines on the surface.

4. With Gray and Black Verithin pencils, add a subtle cast shadow. **(B)**

5. Add a layer of watercolor on top, leaving the highlights empty. **(C)**

6. Continue to layer colored pencils. Close in on the highlights, making them shimmer. **(D)**

7. Burnish with an Ivory colored pencil, add more layers of color, and finish the cast shadow with a touch of Dark Sepia colored pencil.

8. Do another view of your petal—or a different petal if you can't resist! (I couldn't.) **(E)**

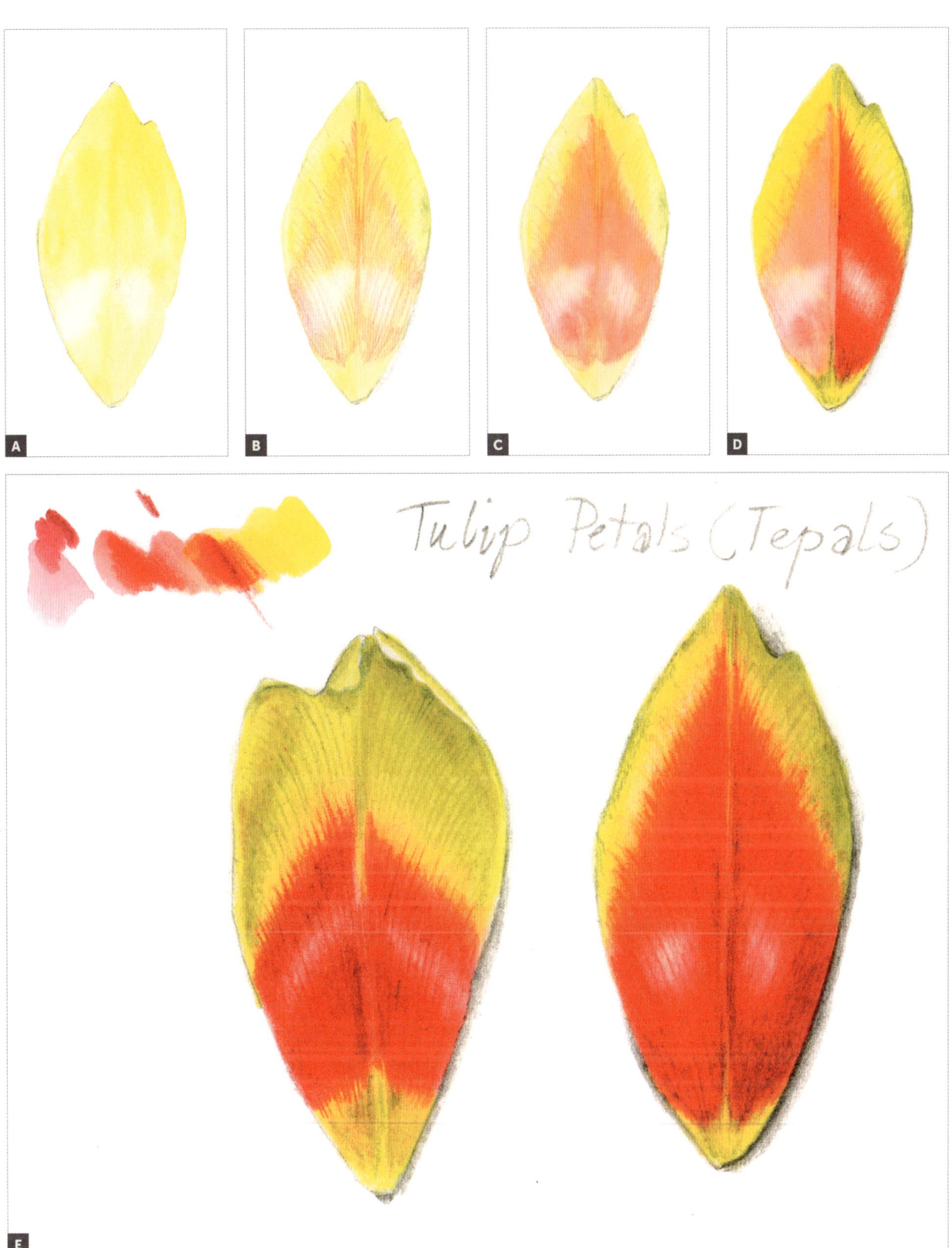

Tulip Petals (Tepals)

Making an Herbarium Page

SUBJECT

Daffodil (which is a combination flower shape) or other flower, such as a tulip, lily, or iris

ADDITIONAL MATERIALS

- Clippers
- X-Acto knife or other dissecting blade
- Magnifying glass or other magnifier
- Scotch tape (optional)
- Plain paper of any kind

Drawing a whole flower can be quite overwhelming, but if you start by observing a flower, taking it apart, and studying the components, drawing an entire flower will be a bit easier. Make sure to refer to the image on page 111 that labels the parts of the flower clearly. It feels manageable to take a flower apart, observe its structure, draw its parts individually, and make color notes. Only after doing this do I gain the confidence to tackle a whole flower. This process is a wonderful learning experience that invites you into the inner workings of nature. Even though I've drawn hundreds of flowers, I still love to start by closely observing a flower before I draw it. I sometimes imagine that the flower appreciates the time I am taking to understand its structure and rewards me by helping me with the drawing. It's a good idea to have several flowers of the same species available to do this exercise so that you'll still have a complete flower intact to draw later on. It's also helpful to have a magnifying glass and a loupe that magnifies up to 30X.

1. Use clippers to cut a few daffodils and take them to your table to work.

2. Use an X-Acto knife to carefully take apart one of the flowers, starting with the petals.

3. Carefully slice open the corona, the center tubular shape of the daffodil. This process will expose the reproductive parts, all of which emanate from the center axis of the stem. You'll find six stamens (male parts), each consisting of an anther (top part) full of pollen and a filament (lower shaft). Lay them on your paper and tape them down.

4. In the center is the pistil (female part) that consists of the stigma (which has three lobes) at the top and the ovary (which has three chambers) at the bottom. Examine the reproductive parts of your daffodil through a magnifying glass. It's easier to see details once you've separated the parts.

5. You can dissect the ovary, which is located below the petals on the daffodil. It is slightly rounded and creates a bulge in the stem. This is called an inferior ovary. There are two ways to dissect the ovary: vertically or horizontally. Slice the ovary in half vertically to reveal the rows of ovules, horizontally to reveal the three carpels or chambers of the ovary. You can see different things about the structure of the ovary depending on which dissection you choose.

6. Arrange all the parts on a plain piece of drawing paper. Label the parts, and then either tape them down or leave them loose. To document the flower's structure, include a whole flower and some leaves. You can also tape flat parts, such as petals, into your drawing sketchbook.

7. Cover the specimen with another piece of paper. Press it under some heavy books or use a press made for this purpose. If you're pressing your specimen right in your sketchbook, put a piece of plain paper over the specimen and then put the closed sketchbook under some heavy books. Your herbarium page will be dry in about two weeks.

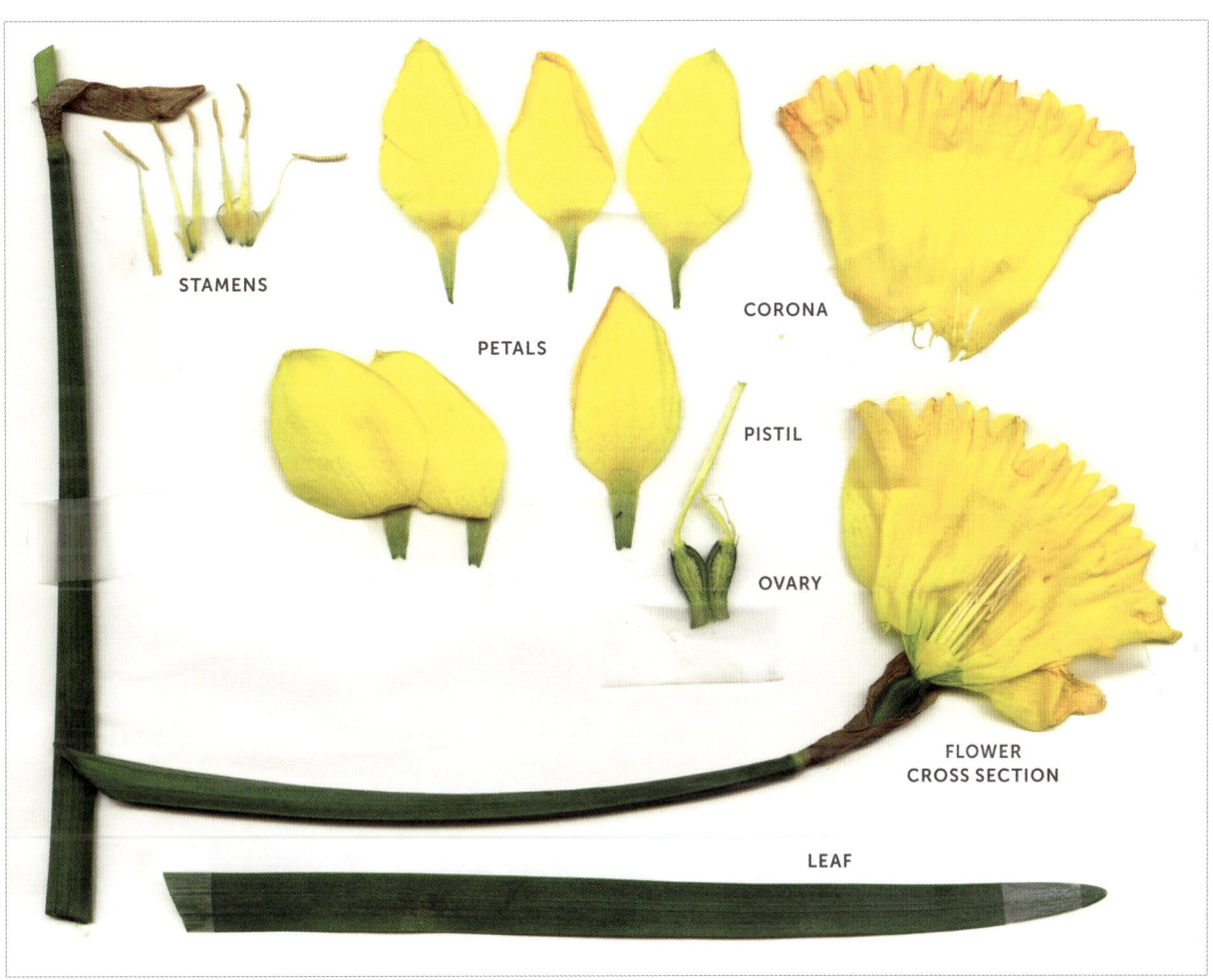

STAMENS

PETALS

CORONA

PISTIL

OVARY

FLOWER
CROSS SECTION

LEAF

Making a Process Page from Your Herbarium Page

SUBJECT

Your pressed flower from the previous lesson

ADDITIONAL MATERIALS

- Your herbarium page from the previous lesson
- Colored pencils and watercolor pencils to match the local color of your flower
- Earth Green colored pencil #172
- Watercolor brush #2
- Magnifying glass or other magnifier

COLORED PENCILS

#172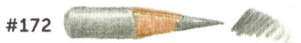

The goal of this lesson is to make a study page with drawings of the separated parts from your herbarium page, and to practice color mixing. These exploratory pages of dissected flower drawings are wonderful to look at, and they allow the viewer to see what you have experienced and discovered. Have a botany book available for reference or use online research to help with labeling and understanding the flower structure. On your page, consider recording the number of petals and details about the reproductive parts and leaves.

You can tape parts onto your process pages. These elements add visual interest to a page. Simple flowers with large reproductive parts—such as daffodils, tulips, lilies, irises, and anemones—are great, because you can easily observe the flower's structure.

Follow the same procedure in this lesson with any other flower types.

1. To study a petal, put one right on the paper next to where you mix your colors as a simple way to match the hue. Draw one individual petal life-size to understand the structure, and practice color blending so that when you start rendering a full flower, you'll know what colors to use. Blend the shadow color as well on the petal so that you work out all your color options now. **(A)**

2. To draw the whole dissected flower, complete an outline drawing and tone overlaps with Earth Green colored pencil, taking special care to tone behind all the reproductive parts. **(B)**

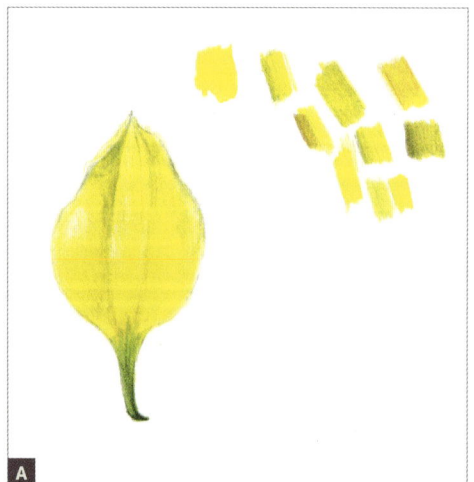

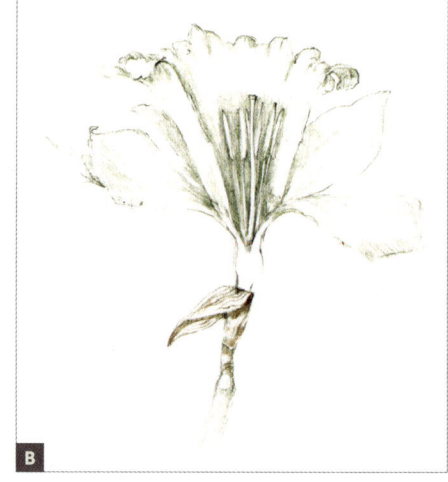

3. Use a #2 brush to add a watercolor wash behind the reproductive parts on the flower, leaving some highlights with graduated yellow watercolor. **(C)**

4. Add details on the flower and stem with colored pencils. **(D)**

5. Draw other parts of the flower on your page. Look at your dissected ovary from your herbarium page using a magnifying glass. Draw the ovary with the ovules inside. Then draw the reproductive parts you see.

6. Draw any other parts of the flower that intrigue you, such as close-ups of leaves, stamens, or pistils. **(E)**

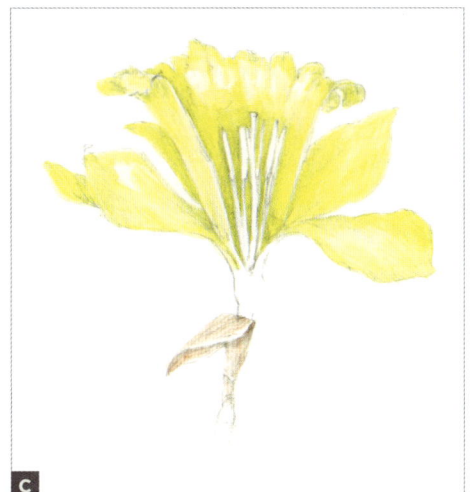

C

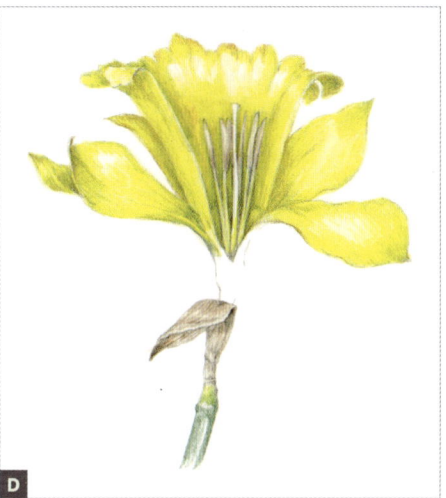

D

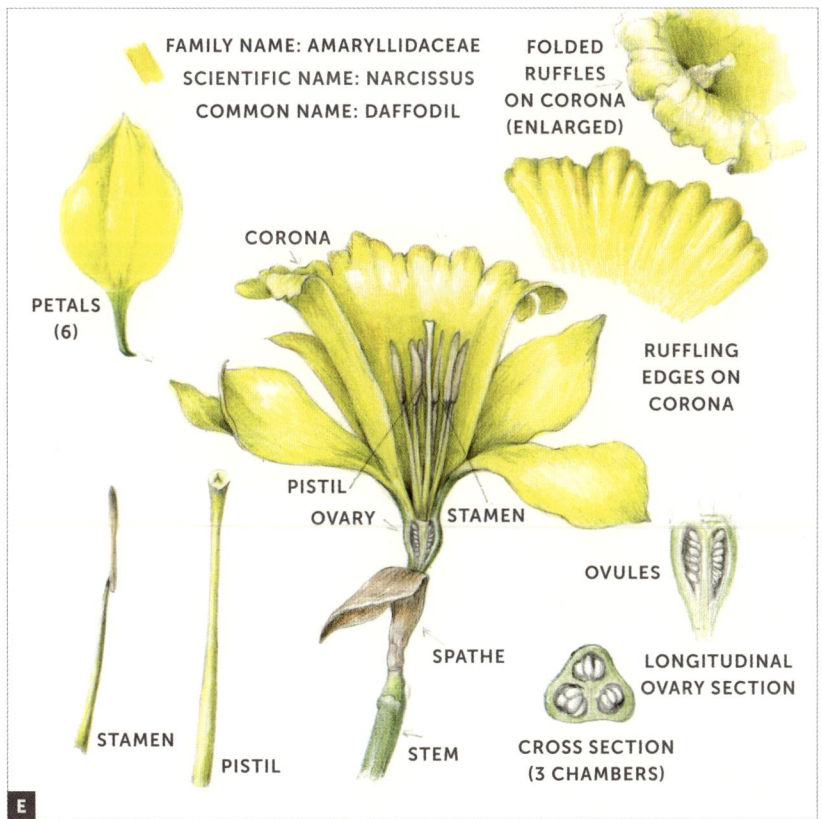

FAMILY NAME: AMARYLLIDACEAE
SCIENTIFIC NAME: NARCISSUS
COMMON NAME: DAFFODIL

FOLDED RUFFLES ON CORONA (ENLARGED)

PETALS (6)

CORONA

RUFFLING EDGES ON CORONA

PISTIL
OVARY
STAMEN

OVULES

LONGITUDINAL OVARY SECTION

SPATHE

STAMEN

PISTIL

STEM

CROSS SECTION (3 CHAMBERS)

E

Drawing a Morning Glory (Tubular Flower)

SUBJECT

Morning glory

ADDITIONAL MATERIALS

- Frog prong or smaller jar (optional)
- Dark Sepia colored pencil #175
- Tracing paper
- Watercolor brush #2
- Colored pencils and watercolor pencils to match the local color of your flower

COLORED PENCILS

#175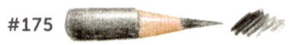

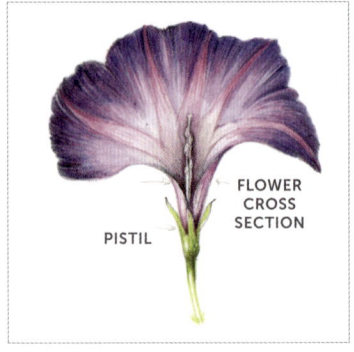

FLOWER CROSS SECTION

PISTIL

I love to draw tubular flowers because the structure can look dramatic and three-dimensional. To do this, I choose a view that shows the tube shape as it opens into a rounded flower, with emphasis on the deep, mysterious center that disappears into the tubular shape. While I was observing these flowers growing on my porch, I watched as a bee flew inside a morning glory and wiggled down into the center of it, down into the tube, and stayed inside the flower for quite a while. The bee then had to backstep its way out of the flower. I was so excited to discover this! I always encourage you to study your plants in the environment as they grow whenever possible. Nature will be alive and active around you, and there's so much to learn from this! For example, if you pay attention to the growing vines of the morning glory, you can find all stages of development, from bud to flower to seedpod, which is a wonderful way to build a composition.

Refer to page 85 for the basics of perspective and measuring. Remember to use ellipses to help measure and understand the three-dimensional structure of a flower. Though I recommend setting up a light source to help you see your shadows and highlights on the real flower, I exaggerate the light and shadows to create the illusion of the three-dimensional structure based on the overlaps and surface contour. At left is a study dissection I made of a morning glory flower as a warm-up before I did this lesson, and I encourage you to make these exploratory pages whenever you have enough flowers and time. Repeat this lesson many times with other tubular flowers. Some of my favorites are trumpet vines, campanulas, and daffodils.

1. Look at your flower from all angles to find a good view to draw. Choose an angle that will show the structure of the flower well and look three-dimensional, rather than a straight-on flat view. I chose a view that shows both the inside and tubular part of the flower. I like to indicate this simple shape with a shaded thumbnail sketch to help me remember my light source correctly. If possible, secure your flower in a frog prong or a small jar with water to prevent it from moving.

2. Measure your flower, and with an H graphite pencil, very lightly draw your flower life-size.

3. Plan the composition of your flower so that you include a leaf and some of the twisting vine-like stems that are characteristic of the morning glory, and draw them in. **(A)**

4. After drawing lightly with graphite pencil, do a total, more precise "redraw" with a Dark Sepia colored pencil. Draw the clear five-petal flower that starts as a tube and then opens with five attached petals in starlike structure, depicting how each petal radiates out of the center of the tube. (Keep your pencil point very sharp, and do not press hard during this precise drawing.) Make note of the variation of line and value on my drawing at this stage. It is not a solid outline, but has various thicknesses and values so as not to appear cartoonlike. **(B)**

5. Place a piece of tracing paper over your drawing and draw the surface contour of the flower on the tracing paper. Use this as a map to help draw and tone the flower with a clear light source. Notice how the surface contour lines all radiate out of the center of the tube. **(C)**

6. Practice tone bars of color, use a #2 brush to mix watercolors, then add dry pencils for shading on your practice dabs.

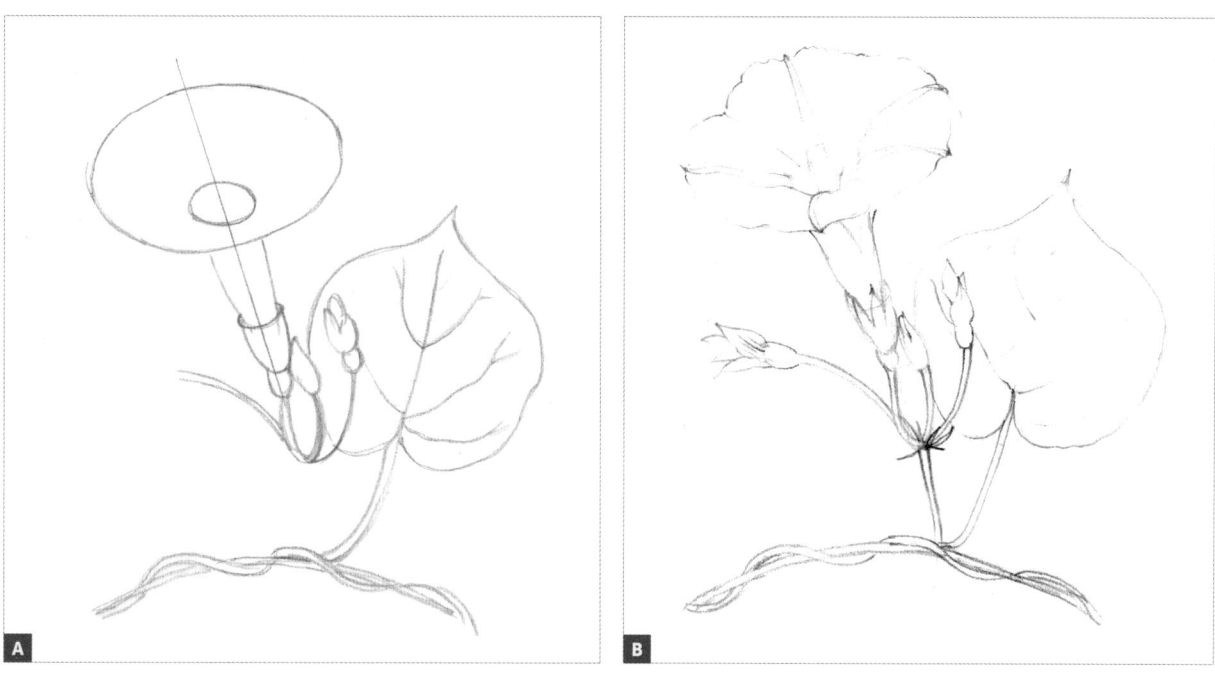

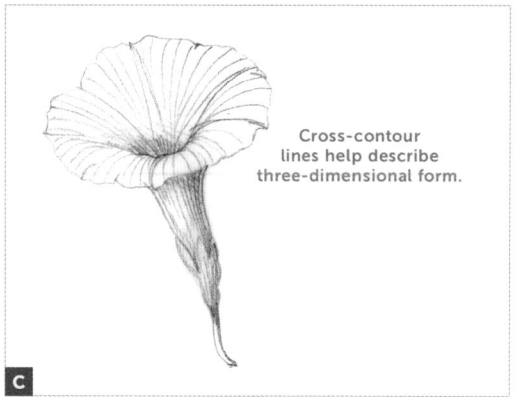

Cross-contour lines help describe three-dimensional form.

CONTINUED

7. Once you're satisfied that you have a selection of pencils that will give the colors you want, paint a layer of watercolor. As this flower has both pink and purple colors, I used both these colors in my watercolor layer. Leave some empty white areas indicating where the highlights will be. Add a layer of green watercolor on the sepal cup, buds, and stems. **(D)**

8. Use colored pencils to add overlaps and color on the flower.

9. With colored pencils, and Dark Sepia on the overlaps, add color and detail to the buds, stems, and leaves. **(E)**

10. The morning glory flower has colored venation that follows the contour lines of the petals. Continue layers of color, using this pattern to define the veining of these contour lines. Layer color on the other parts of the plant as well.

11. You can also add to the composition by drawing a dried seedpod that develops on the vines, and scatter in a few seeds. **(F)**

12. With watercolor, add color on the seedpods and then, once it dries, use colored pencils on top to complete the seedpods. I added in a bumblebee pollinator; do this if you'd like. **(G)**

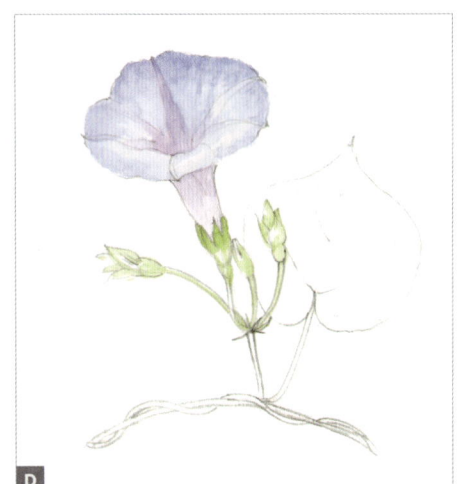

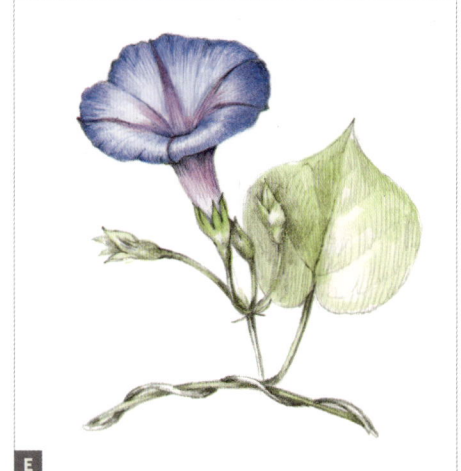

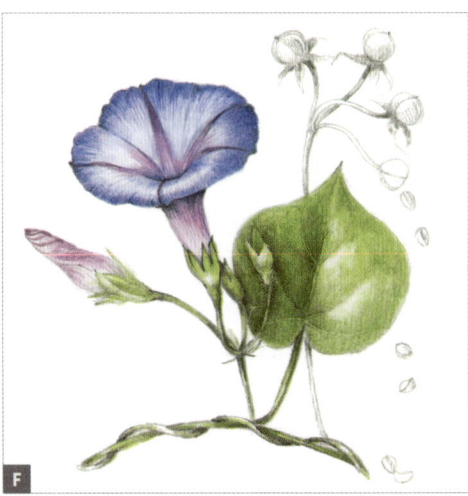

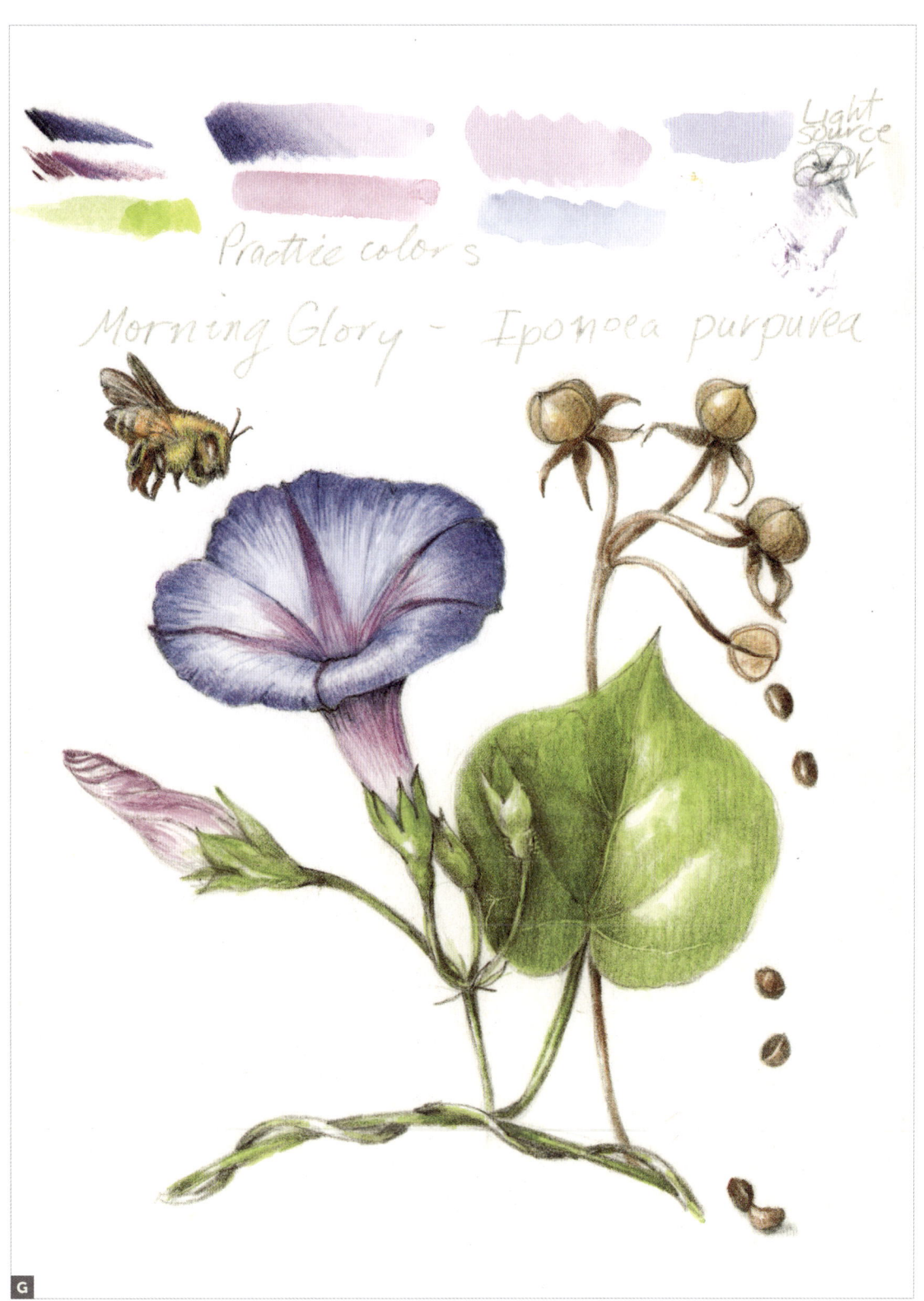

Practice colors

Light source

Morning Glory - Ipomoea purpurea

Drawing an Allamanda (Tubular Flower)

SUBJECT

Allamanda or other tubular flower

ADDITIONAL MATERIALS

• Watercolor brush #2
• Watercolor pencils and colored pencils to match the local color of your flower

By drawing several different flowers that have a tubular shape, you'll gain confidence, and tackling similar flower shapes will become easier. You'll also start to notice the ways that flowers are similar, but also their subtle differences. Delight in the diversity that nature exhibits and highlight it in your drawings! This allamanda is interesting to draw in part because of the beautiful gold color of the tubular part of the flower that changes to bright yellow when the petals split apart and open. Yellow, because it is a light color, is challenging to make look three-dimensional, so be sure to refer back to the color study on page 64 for good shadow colors to use.

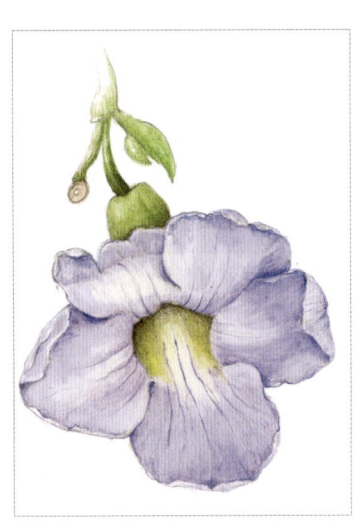

1. With a graphite pencil, draw the flower life-size and notice what simple geometric shape it is similar to. I like to indicate this simple shape with a shaded thumbnail sketch to help me remember my light source correctly.

2. Draw cross-contour lines on the flower, following the veining of the petals. **(A)**

3. Tone the overlaps and then add tone to show the correct light source. **(B)**

4. Using a #2 brush, build up layers of color with a watercolor wash and then layer colored pencil. Emphasize details you see on your subject with toning and different colors. Burnish to blend the layers together, leaving a good highlight where appropriate. Add other elements to your composition, such as leaves and buds.

5. Add more layers to distinguish details. **(C)**

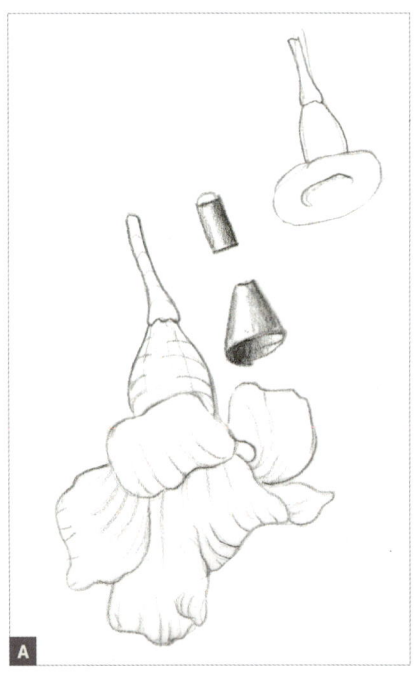

A

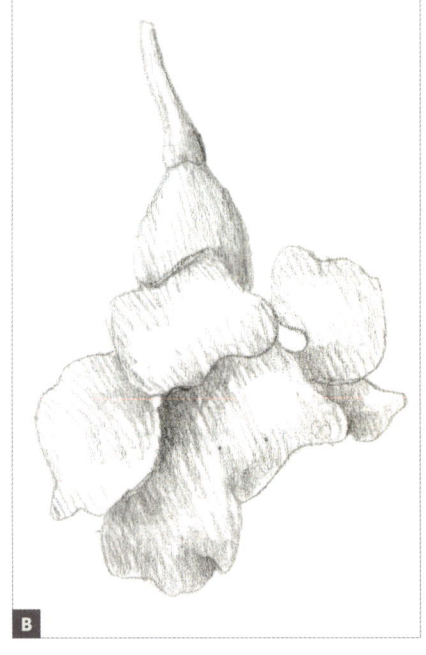

B

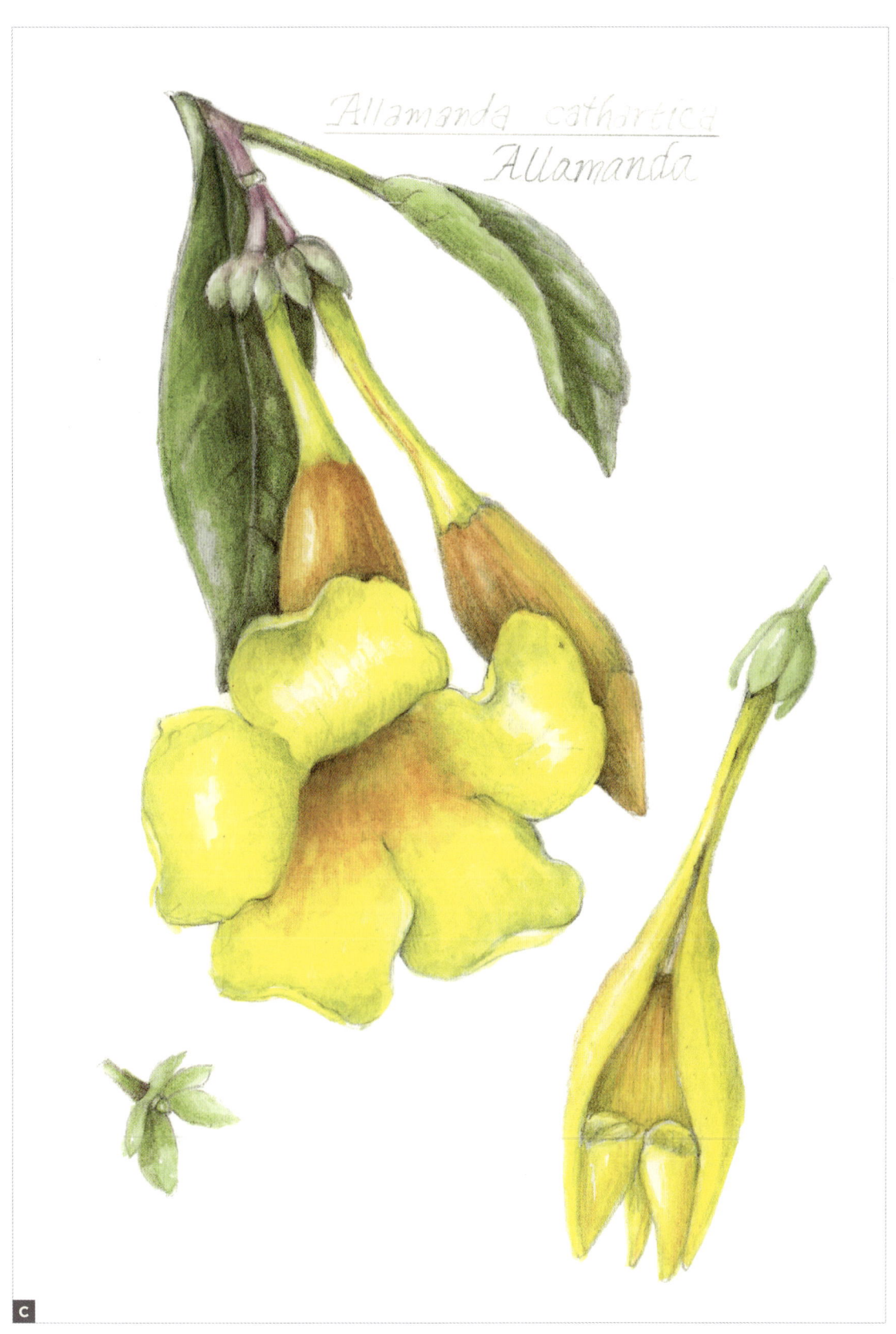

Allamanda cathartica
Allamanda

C

Drawing an Anemone
(White Disk-Shaped Flower)

SUBJECT

Two anemones of the same color

ADDITIONAL MATERIALS

- Tracing paper
- Watercolor brushes #2 and #³/₀
- Watercolor and colored pencils to match the local color of your flower
- Embossing tool (small)
- Warm Grey IV colored pencil #273
- Dark Sepia colored pencil #175

COLORED PENCILS

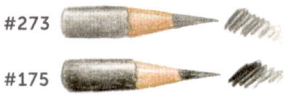

#273

#175

Rotate-shaped and disk-shaped flowers that are mostly flat and circular, such as an anemone, are good practice for drawing many other kinds of flowers. Rotate flowers have petals that all radiate out of a center where the reproductive parts live. When drawing a flower with complex reproductive parts, carefully draw those elements first so that when you start to work on the petals, they'll be attached to the center of the flower. Use ellipses to measure your view in perspective, and make sure to draw each petal with its own center axis, radiating out of the center of the flower. Note that due to foreshortening, the center of the flower isn't always in the center of the ellipse! Because of this, some petals (especially those in front) will appear foreshortened, and this is the secret to making interesting and realistic three-dimensional views.

When I drew this flower, it was the middle of winter and my gardens were dormant; however, about a 45-minute drive from my home is a nursery that grows anemone flowers for cutting all year and ships around the world. I was excited to see these flowers growing in the many hothouses, and I took home several bouquets to work from. Anemones come in so many colors that it's always hard to choose one to draw. I chose white with some lovely purple venation to show you some of the important techniques you can use with white flowers.

1. Begin by setting up your flower so that it is stable and will not move. (I balanced my flower inside a jar with the petals rolling over the jar's edges.) Once you have a view you like, immediately snap some photos of your setup, because anemones (and many flowers) will open and close at various times of the day.

2. Measure the flower and create an ellipse around the petal edges. Locate the center of the flower and, using a graphite pencil, draw a center axis in the direction each petal radiates out of the center. You can measure from one tip of the petal to the next to help get the measurements correct. The foreshortened petal in the front is always hardest to draw. Be sure to measure the width as well as the depth. **(A)**

3. Practice your drawing on tracing paper and add in surface contour lines, following the venation of the petals. Make sure to indicate how the petals overlap and curl. Do a light-source thumbnail sketch to help with grisaille toning. **(B)**

4. Redraw your flower on good paper. Carefully draw the reproductive parts so that when you start to work on the petals, they will appear attached to the center of the flower. **(C)**

5. Take a single petal off of your second flower to study and practice drawing a petal. This way you can practice the colors, veining pattern, and rolling of a petal before tackling the whole flower. **(D)**

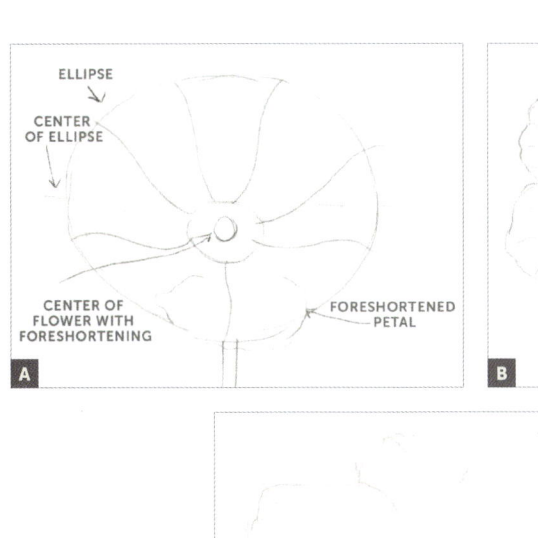

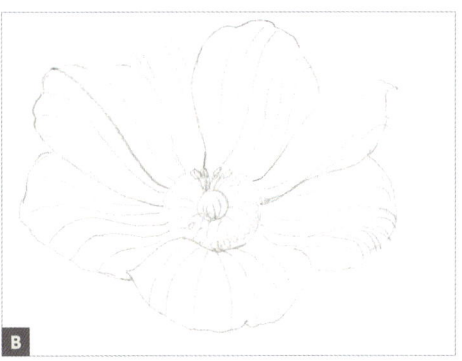

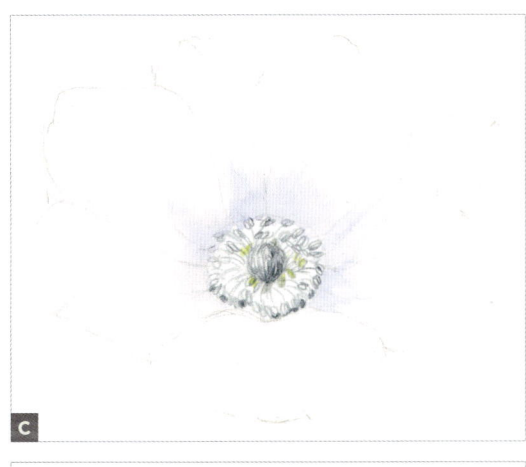

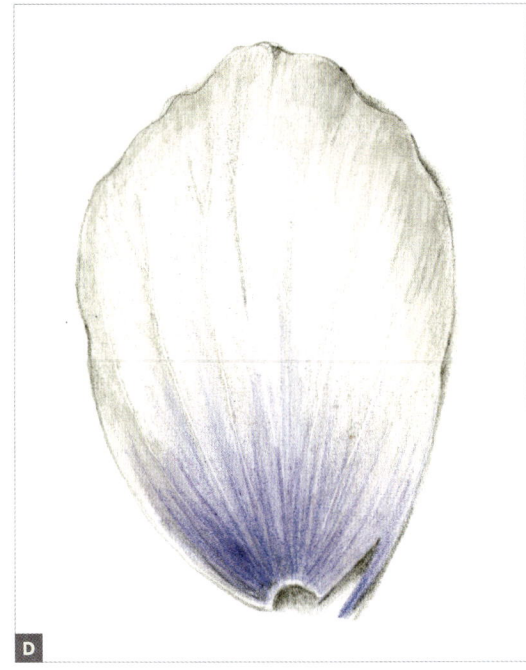

CONTINUED

6. With a #2 brush, create a watercolor wash and add some light watercolor to each petal to describe the petal venation, using a mixture of violet and blue to create the exact shade of blue-violet. With an embossing tool, draw some fine veins. With a light shadow color, such as Warm Grey IV colored pencil, tone the flower lightly. **(E)**

7. With colored pencil, continue to work on the detail in the center of the flower, and use the embossing tool to maintain light areas, such as the filaments on the stamen. **(F)**

8. Add color to the venation of the petals and more layers of the white petals, leaving good highlights and a lot of light, almost white, so that the flower appears white and not gray.

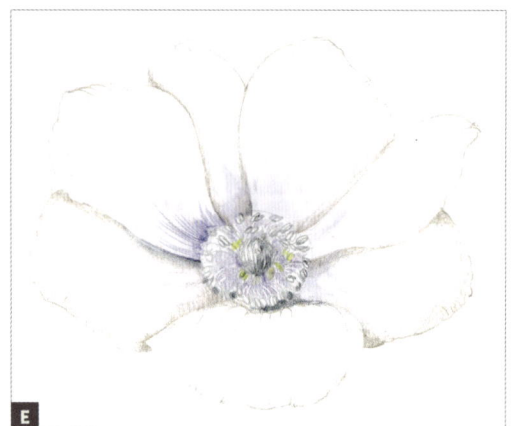
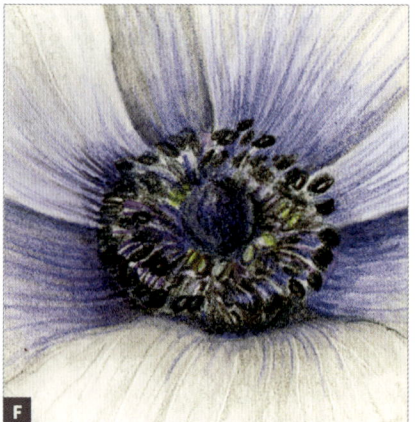

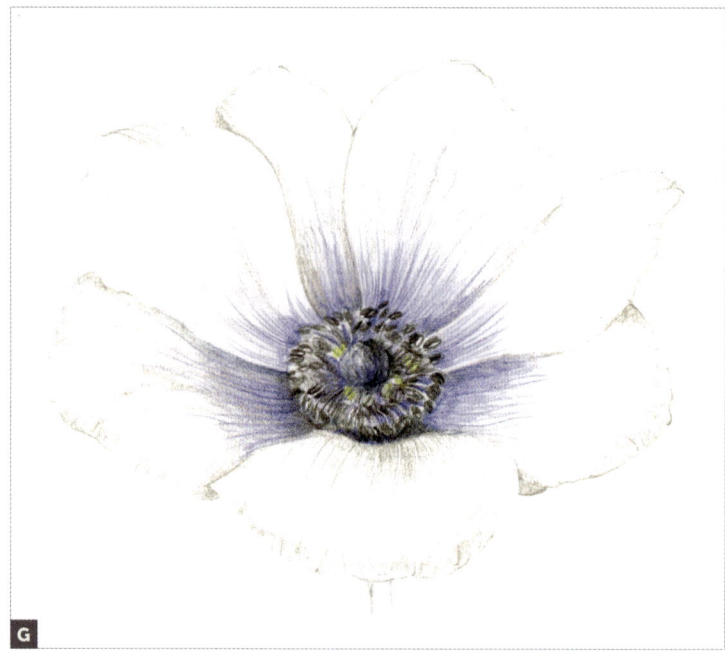

9. With gray watercolor, gray colored pencil, and just the slightest amount of Dark Sepia colored pencil, draw subtle, delicate inside edges and curling petal edges.

10. To fill in the center of this flower that is very detailed and dark, use an ³/₀ fine-point brush with dry-brush pigment mixed to match the flower color. **(G)**

11. Finish with the stem and the lovely curling leaves of the anemone. **(H)**

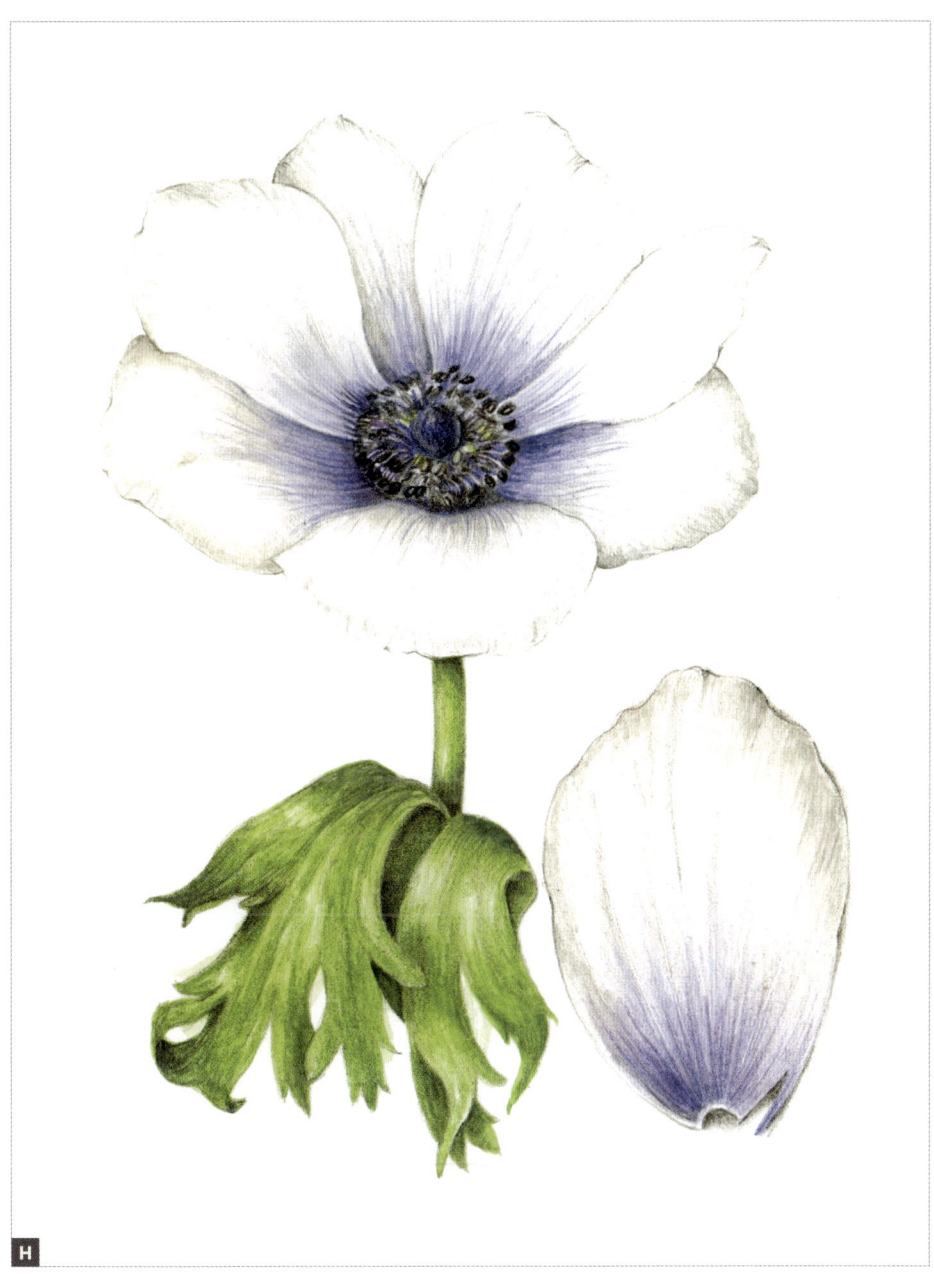

Drawing a Zinnia (Composite Flower)

SUBJECT

Zinnia or other composite flower, such as a daisy or sunflower

ADDITIONAL MATERIALS

- Magnifying glass or other magnifier
- Watercolor brush #2
- Colored pencils and watercolor pencils to match the local color of your flower
- Embossing tool (small)

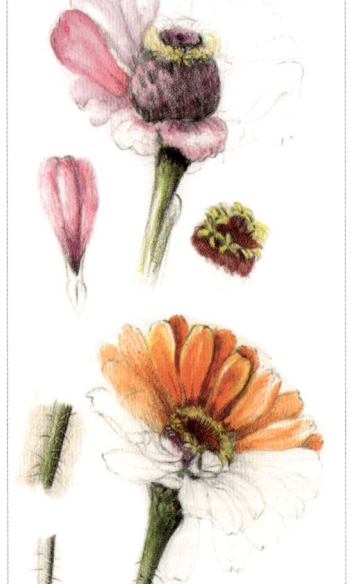

In my garden, we plant fields of these colorful flowers each summer. They last well into the fall when most other flowers are on their way out, and monarch butterflies love them and dance from one bloom to the next.

Composite flowers are interesting and fun to draw, but you should avoid drawing a flat circular view of this kind of flower and instead go more for a foreshortened elliptical view because it will look more three-dimensional and thus show structure better. The structure of this family of flowers is interesting, as each flower is actually lots of flowers on a stem. Each petal is technically a flower, and the center of the flower is made up of many tiny flowers as well. The flowers in the center contain the female reproductive parts, and this is where the seeds develop. I used to draw the center of composite flowers, such as a daisy, as a round flat color, but no more. I love to show and explore the dimension and details of the flower centers. In addition to this step-by-step lesson, look at my process drawings of zinnias on page 155. They show some of the unusual characteristics and structural challenges in drawing this flower. The key is to render the center of the flower extremely three-dimensional and detailed. Make sure to study under magnification to see and draw these details.

1. Measure your flower and draw an elliptical view of it. Draw your center axis radiating from the center of the flower through the stem. Draw the center axis of each petal radiating out of the center of the flower. **(A)**

2. Look at the center of the flower under a magnifying glass. Draw the center of the flower and then start drawing the petals, beginning with the petals on top. **(B)**

3. Add petals underneath, or behind, the top petals, and tone with overlaps so that you're clearly describing the layering of the petals. **(C)**

4. Practice color mixing and then, with a #2 brush, add a watercolor wash to the petals and the flower's center. Make sure to render the center in a three-dimensional way with good grisaille toning, fine details using an embossing tool, and good contrast.

5. Add in the stem and leaves with grisaille toning and color, and be sure to pay attention to how the leaves attach to the stem as they wrap around it, creating a pleated surface. **(D)**

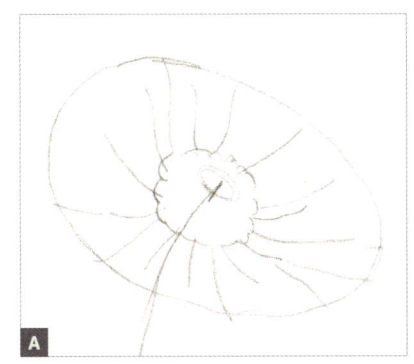

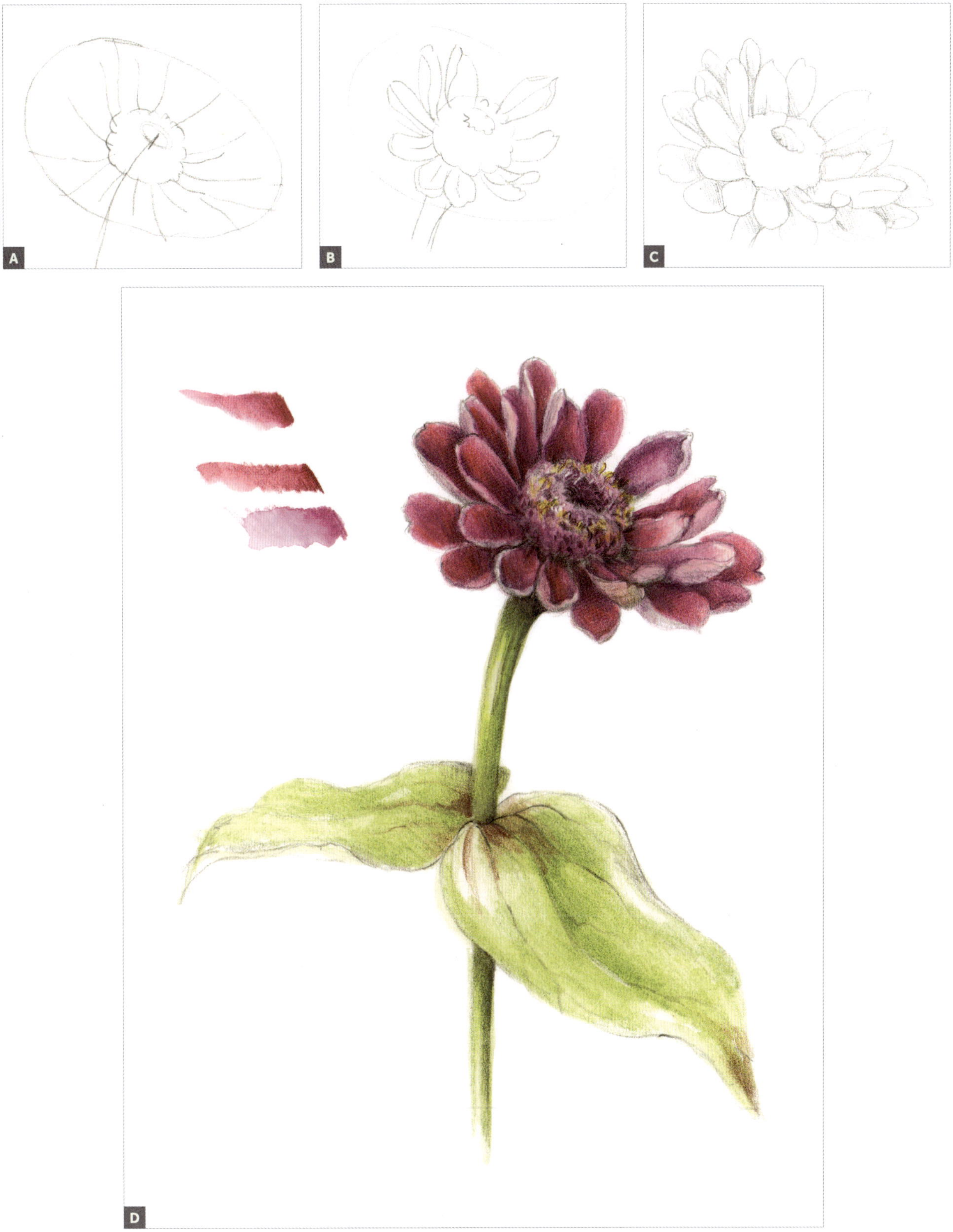

A

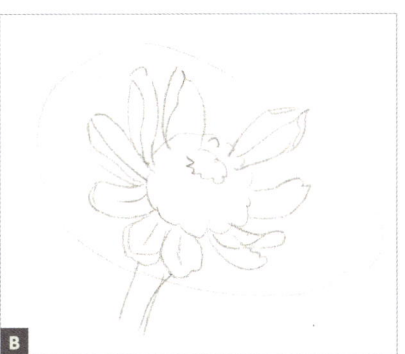

B

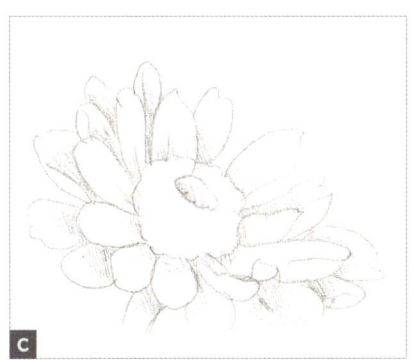

C

D

Drawing a Hibiscus
(Disk- or Funnel-Shaped Flower)

SUBJECT

Hibiscus or other colorful flower

Flowers in the mallow family, such as hibiscus, grow in all kinds of climates. We think of hibiscus as the iconic tropical flower, but they have many temperate-climate relatives. A few fun facts about the hibiscus: Each bloom lasts only one day, but more are always ready to open. The vegetable okra is in the hibiscus family, and its flower is a gorgeous hibiscus. The petals are nicely undulating folds, so always fun to draw, and the center reproductive parts are housed on a dramatic column (think cylinder). Sometimes I like to leave my drawings unfinished, as I have in this case, doing grisaille toning with a Dark Sepia colored pencil on the stem and leaves, but not adding in color. This is a technique that invites the viewer into the page and your process, and lets them focus on the flower itself.

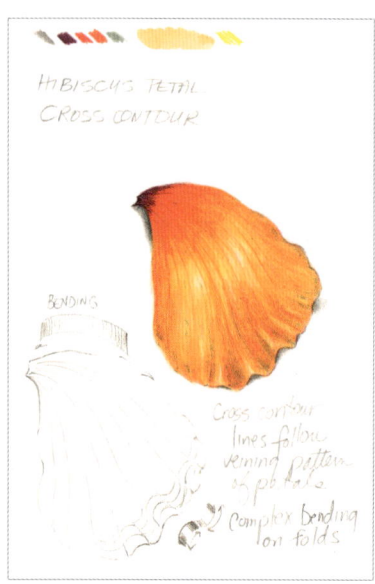
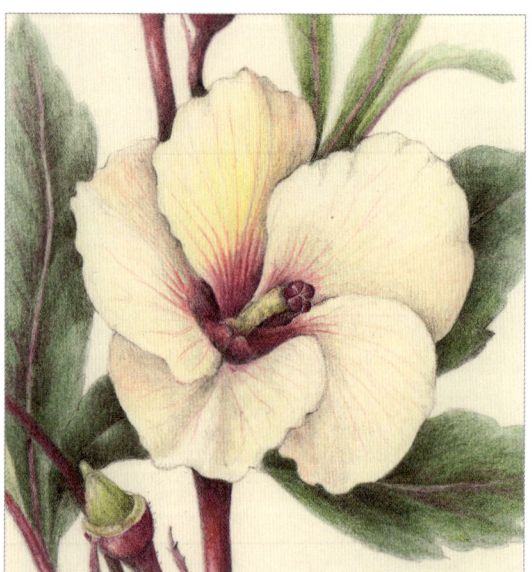

1. Draw a flower bud as a nice warm-up, practicing grisaille toning and layering color. **(A)**

2. Draw a petal of a colorful flower, and really enjoy drawing the variety of the surface contours and vibrant colors. **(B)**

3. Draw the entire flower, making sure to choose a three-dimensional, slightly elliptical view, and start with a good drawing of the center reproductive column. Make sure the petals radiate out from the center.

4. Finish your drawing by continuing to add layers of local colored pencil. Emphasize details you see on your subject with toning and different colors. Burnish to blend the layers together, leaving a good highlight where appropriate. Add more layers to distinguish details. **(C)**

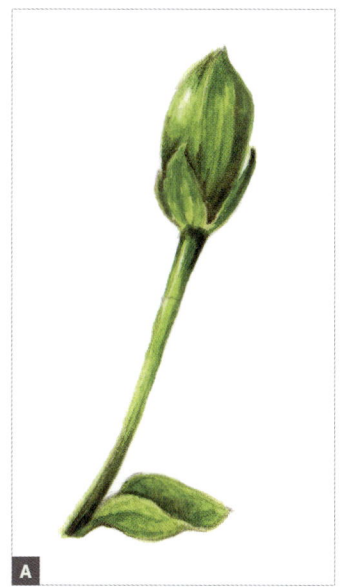

A

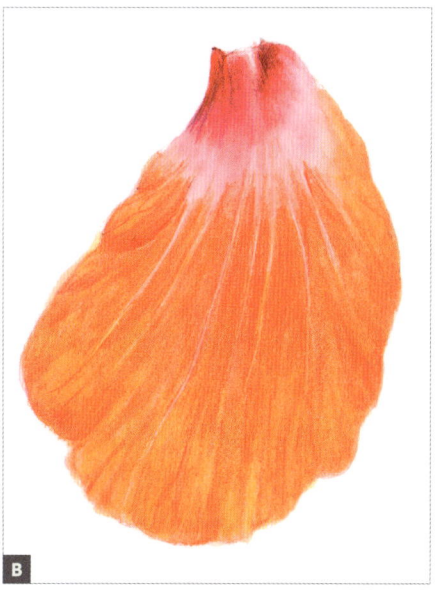

B

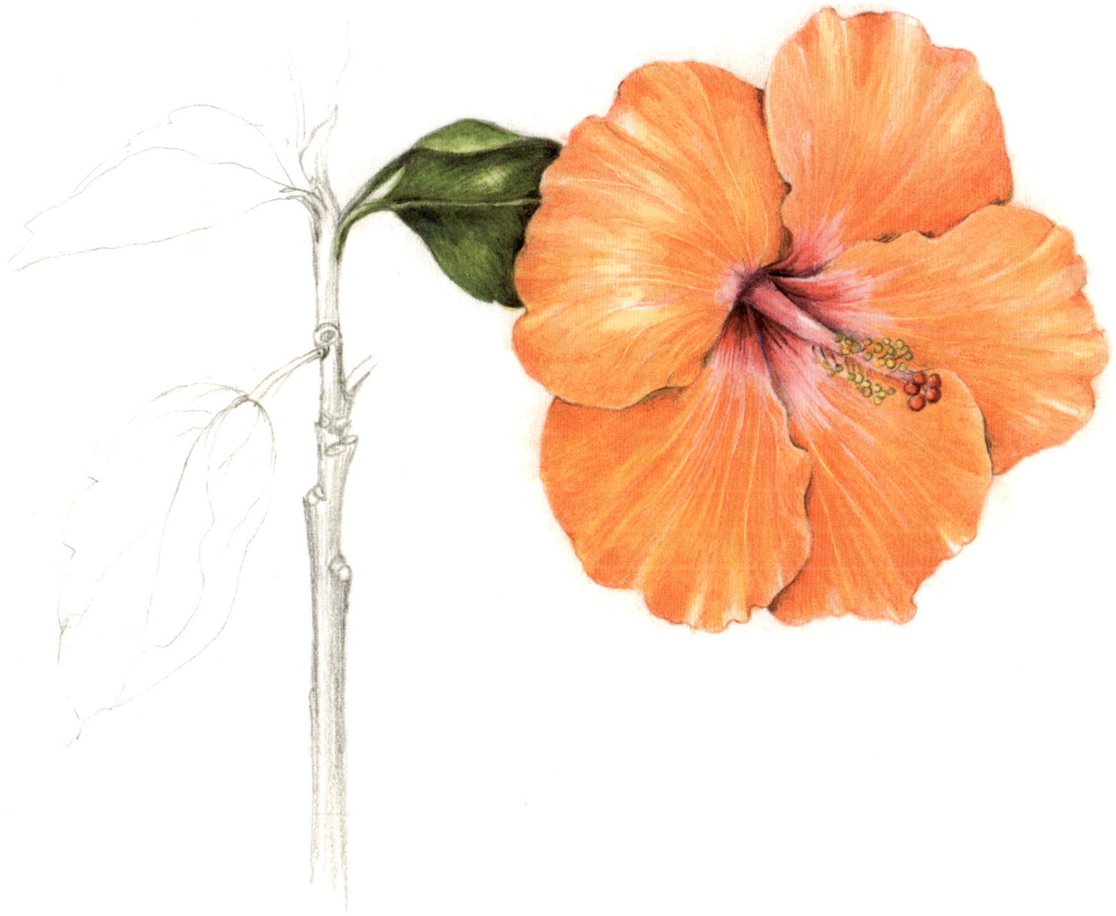

C

Drawing a Rose (Cup Flower)

SUBJECT

Rose

ADDITIONAL MATERIALS

- Tracing paper
- Watercolor pencils and colored pencils to match the local color of your subject

I owe my love of botanical illustration to the romantic beauty of old paintings of cabbage roses often seen in antique chintz fabric designs and prints. They are quite enchanting, in part due to the use of dramatic light and shadow.

Roses with many tightly packed petals can be challenging to draw. Often there are so many petals that drawing this subject can be dizzying.

Notice the overall cup shape to the flower. I consider the overall form of a rose and the way the petals hug the form, and then how they peel away in bending shapes that can be rendered as cylinders. I pay close attention to the details of overlapping petals and make sure to describe this in the beginning of my drawing.

To start, focus on the overall shape of the whole flower. Set up your subject to give a good view of a rose blossom with lots of contrasting dark areas of shadow versus areas highlighted by a light source. The shiny leaves have planes that bend, creating dramatic highlights and shadows.

Get ready to spend several hours or even days on a drawing. Since a rose can keep opening as you work, putting your rose in the refrigerator can help preserve it. If you find yourself getting frustrated and impatient, stop to take breaks. Remember to "smell the roses" as you draw and study!

1. Draw a thumbnail sketch of your rose for overall form and light source. It is helpful to do a few loose concept sketches of various perspectives to make sure you have a dramatic light source and pleasing view. Snap some photos for your reference. **(A)**

2. Measure your rose and, with a graphite pencil, draw each petal carefully but lightly. Notice how the petals appear to be rolling, and that there is an overall cup shape to the flower. Place a small piece of tracing paper over the drawing and draw cross-contour lines to describe the bending, rolling petals for reference. **(B)**

3. Apply a layer of grisaille toning, emphasizing the overlapping petals with a good light source. If your flower is a pale color, start with light toning to keep your colors fresh. **(C)**

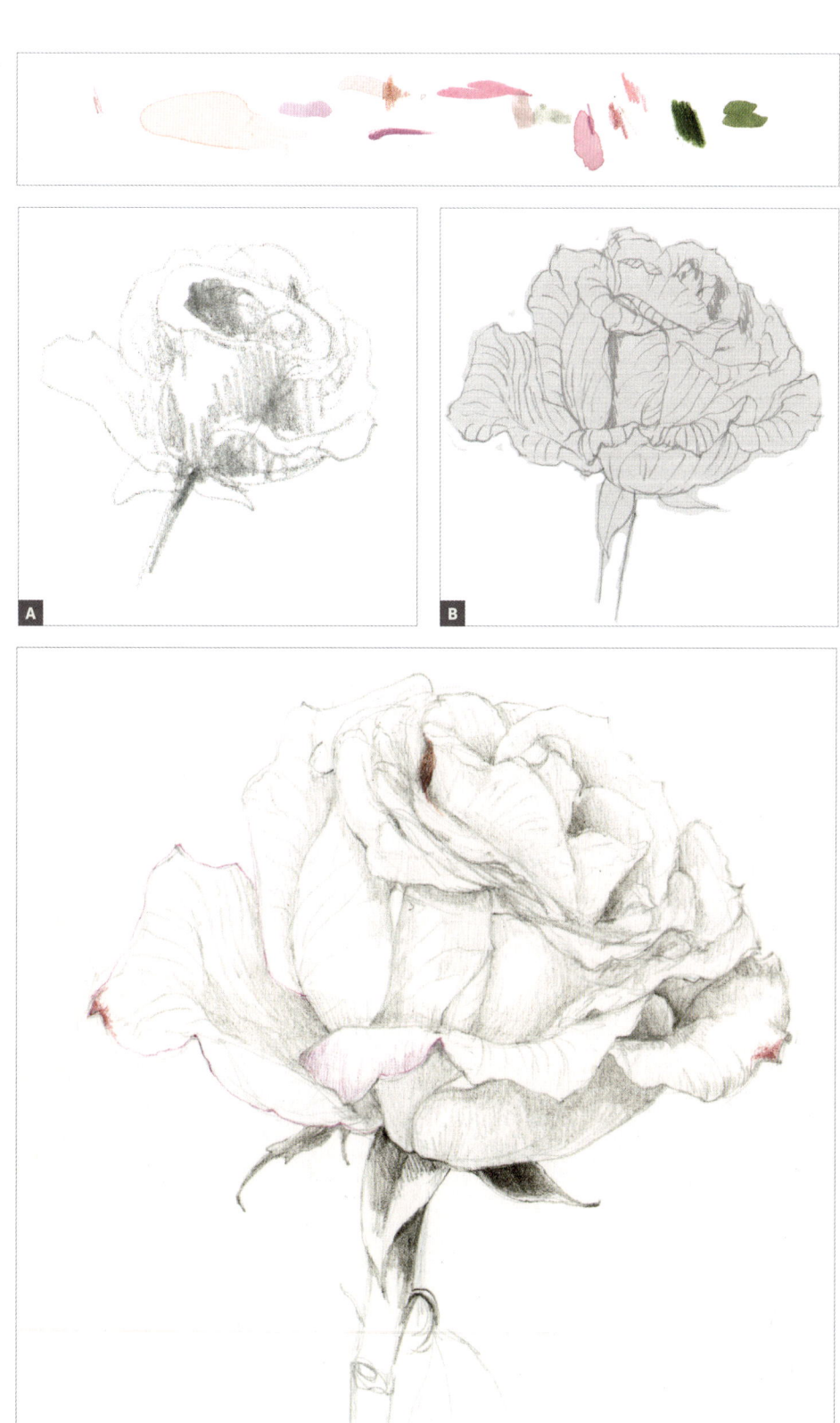

CONTINUED

4. Add a layer of watercolor to the flower, leaving the highlights as the white of the paper. **(D)**

5. Continue to layer color, maintaining a strong contrast between the light and dark areas. Emphasize the shape of the petals and direction in which they roll with toning. Add details and sharpen edges.

6. Create a compelling focal point by contrasting the pale flower with dark, shiny, serrated leaves. Leaves have serrated margins and are often dark with shiny highlights, a good way to create contrast around a pale flower. **(E)**

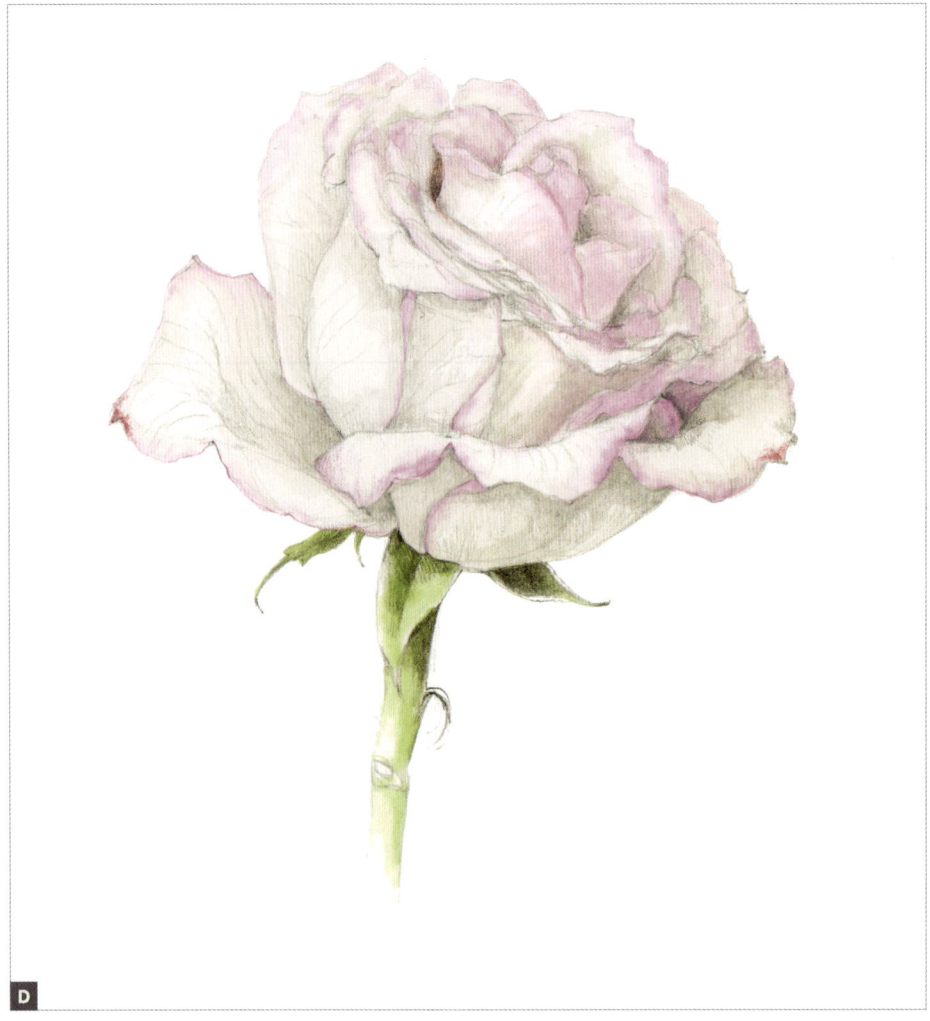

D

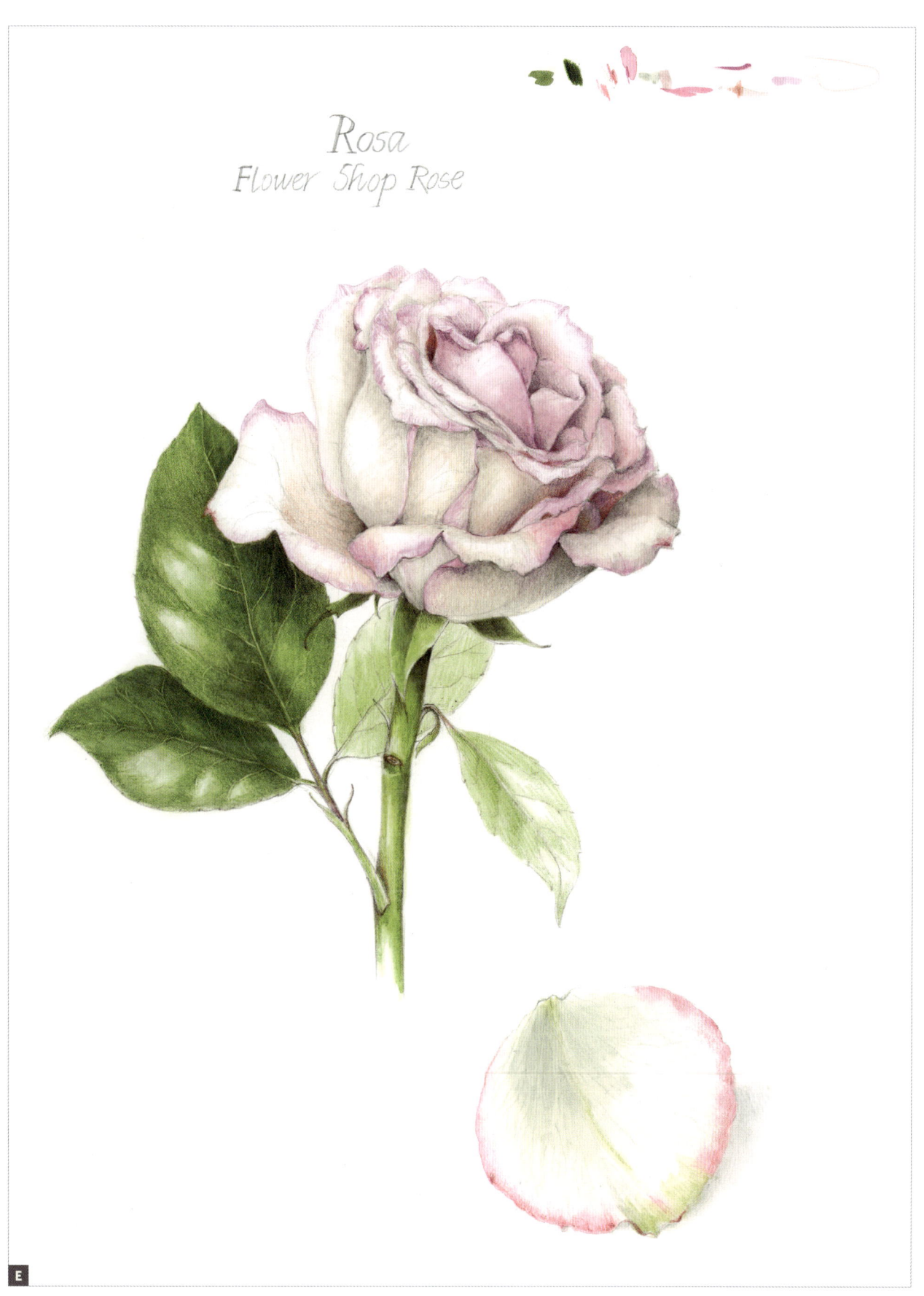

Rosa
Flower Shop Rose

APPLE - "Firiki from Pelion", Greece

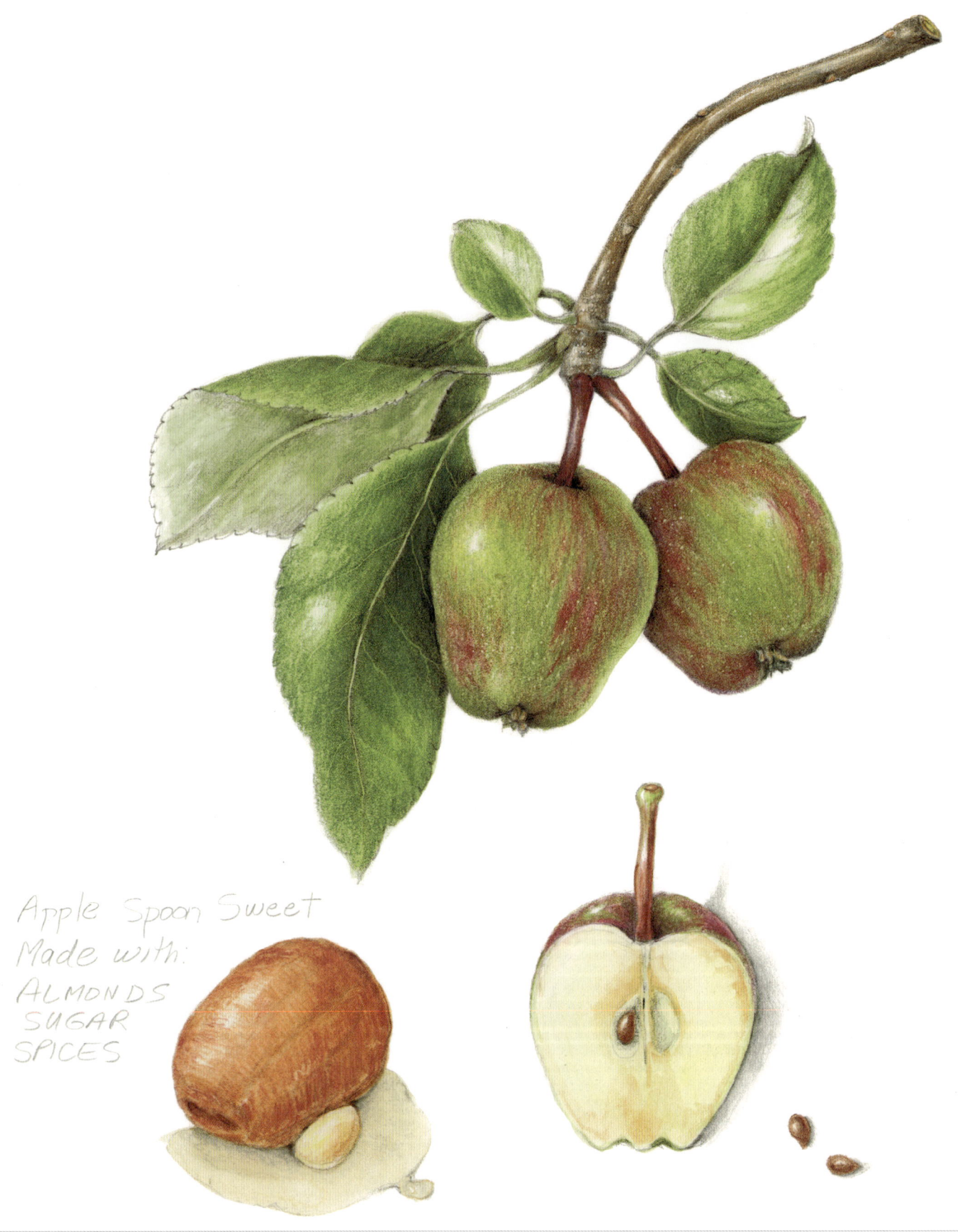

Apple Spoon Sweet
Made with:
ALMONDS
SUGAR
SPICES

APPLE - "Firiki from Pelion", Greece

CHAPTER EIGHT

CROSS-CONTOUR AND PATTERNS ON FORM AND COLOR

When we study nature closely, the same patterns appear again and again. Spiral patterns are clearly visible throughout nature, including in animals and shells. In plants we see them on an acorn's cupule, a pinecone, a pineapple, a strawberry, and on composite flowers, such as the center of a sunflower. It's reassuring to me to witness nature's organized way of creating the structure of all living forms. I invite you to look and notice nature's patterns wherever you go; once you're a visual observer of nature, you'll start to see the same patterns over and over. Spirals, veining, tessellation, and color venation all use repeated structures. Sometimes cross-contour lines will help you describe the correct pattern on three-dimensional form.

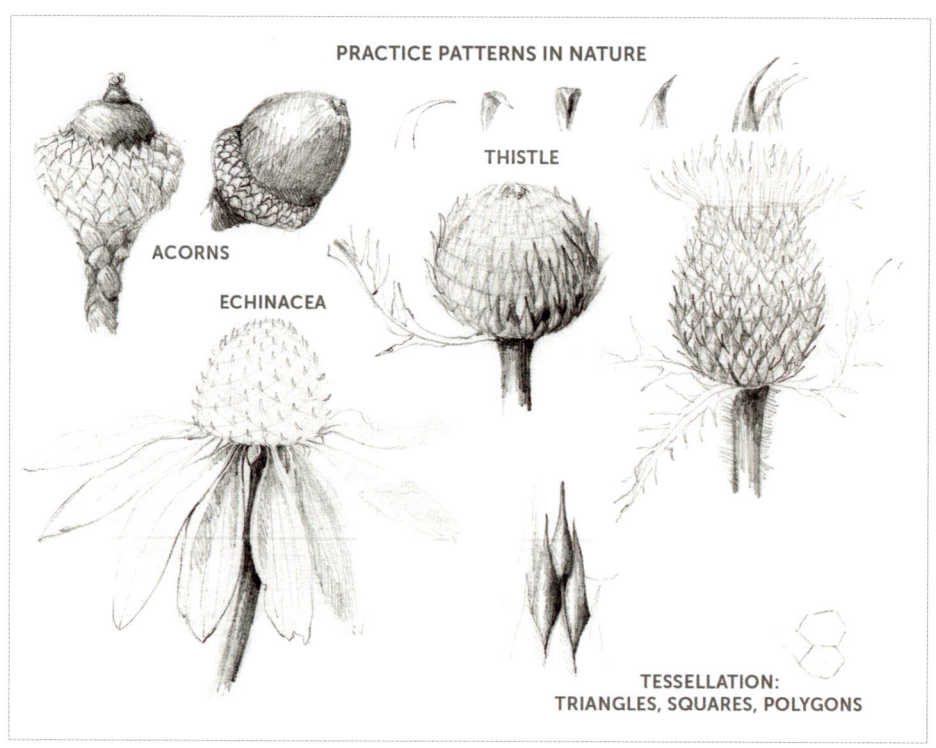

PRACTICE PATTERNS IN NATURE

ACORNS

THISTLE

ECHINACEA

TESSELLATION:
TRIANGLES, SQUARES, POLYGONS

Cross-Contour and Pattern on an Apple

SUBJECT

Apple with strong color patterning

ADDITIONAL MATERIALS

- Red Violet colored pencil #194
- Colored pencils and watercolor pencils to match the local color of your apple
- Watercolor brush #2
- Embossing tool (medium)
- Ivory colored pencil #103
- Brown Verithin pencil
- Black Verithin pencil
- Cool Grey Verithin 70% pencil

COLORED PENCILS

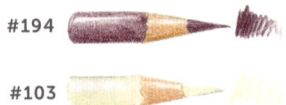

#194

#103

Sometimes apples have contrasting color variation with a definite pattern that conforms to the surface contour of the apple. The pattern appears to follow cross-contour lines. This is an opportunity to use cross-contour lines to help describe the pattern and color of the apple, and as a result, your drawing will look very three-dimensional as long as you make sure to use good grisaille toning as well. The tricky part is to continue shading shadows, leaving a good highlight and creating a reflective highlight at the same time. Remember, even though creating a convincing pattern is important, the most important part of the drawing is to convey the three-dimensional quality of the apple. Many subjects in nature have similar color variation and venation that often follow the surface contour.

1. Choose a view of the apple that shows the front, but also reveals where the stem attaches. With a graphite pencil, draw a life-size outline of your apple. Notice where the stem appears and mark a dot as the center point where it is attached to the apple.

2. Looking at the apple from above, its shape is a circle, but for our view, we're tilting the apple so that we see more of the front of the apple. With the tilt, what was a circle now becomes an ellipse. Use your finger to follow the lines of the form and trace the three-dimensional surface contour. To create the illusion of the recessed area where the stem attaches, draw cross-contour lines emanating from the center point where you drew the stem dot. Notice how the apple is similar to a sphere with one difference: there is a recessed area in the center of the top of the apple where the stem attaches. **(A)**

3. With Red Violet colored pencil, lightly indicate cross-contour lines. Add grisaille toning of overlaps next to the stem and then shading on the shadow side of the apple. **(B)**

4. Using a #2 brush, mix up a watercolor wash for a base coat. I created a light yellow-orange underlying color. Paint the apple with a graduated watercolor wash, leaving a good, blank highlight. Also paint a swatch of watercolor wash to use as a practice area for the apple's pattern and color. **(C)**

5. When the watercolor is dry, practice embossing for the dots on the practice swatch, and layer colored pencils to copy the pattern and color of your apple. **(D)**

6. When you're satisfied with these results, emboss the dots on your apple drawing and start layering colors, following the cross-contour lines of the apple. Use the pattern on your apple as a guide, and try to copy what it looks like. **(E)**

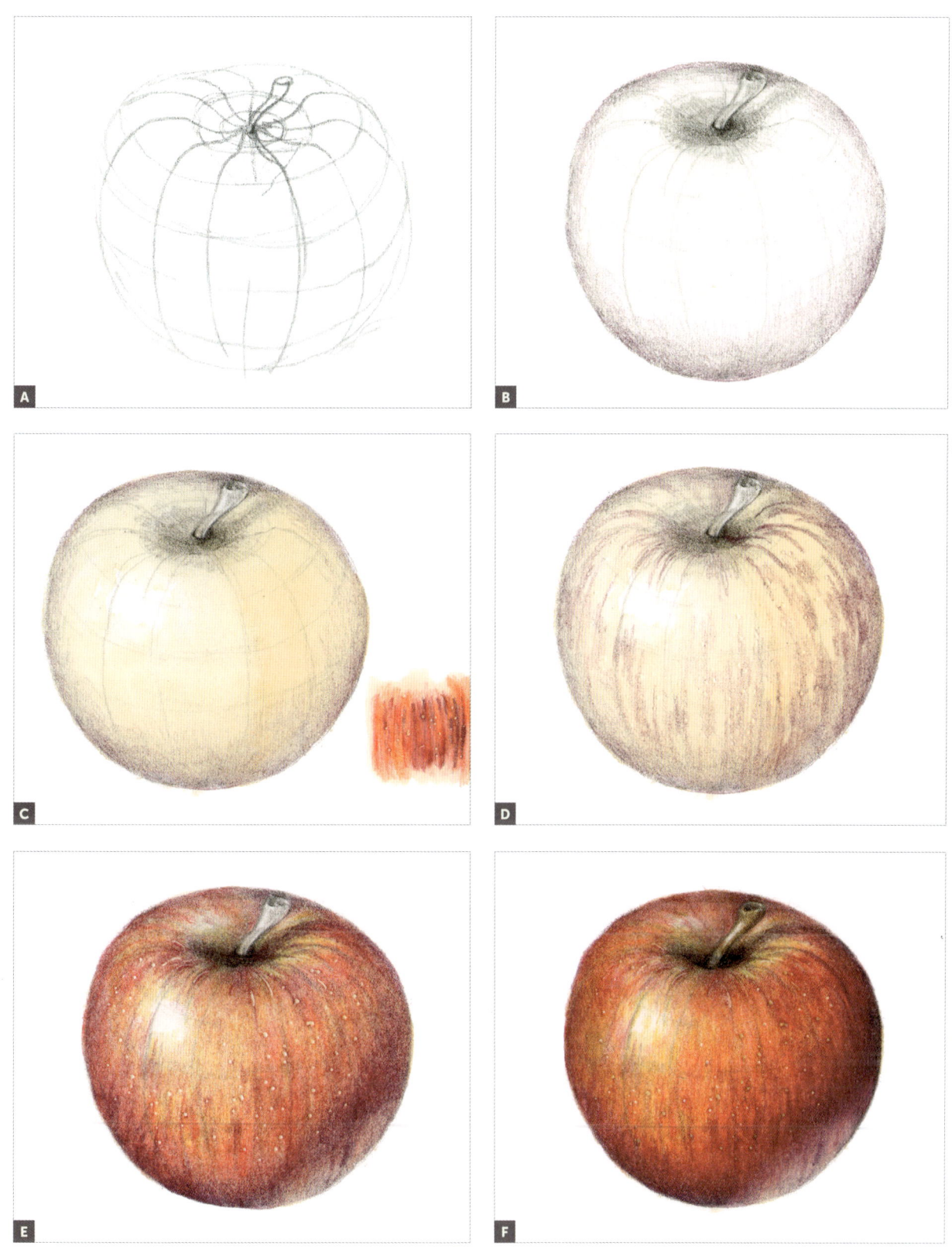

CONTINUED

7. Tone the stem of the apple as you would a cylinder and leave a good highlight.

8. Refine the details by burnishing with an Ivory colored pencil, then fill in edges with Brown, Black, and Cool Grey 70% Verithin pencils to crisp up the contour edge of the apple between the stem and the recessed area. **(F)**

9. As a bonus, show some other elements of your apple: a section of peel, seeds, whatever strikes you as fun to add! Add details and cast shadows with the Verithin pencils. **(G)**

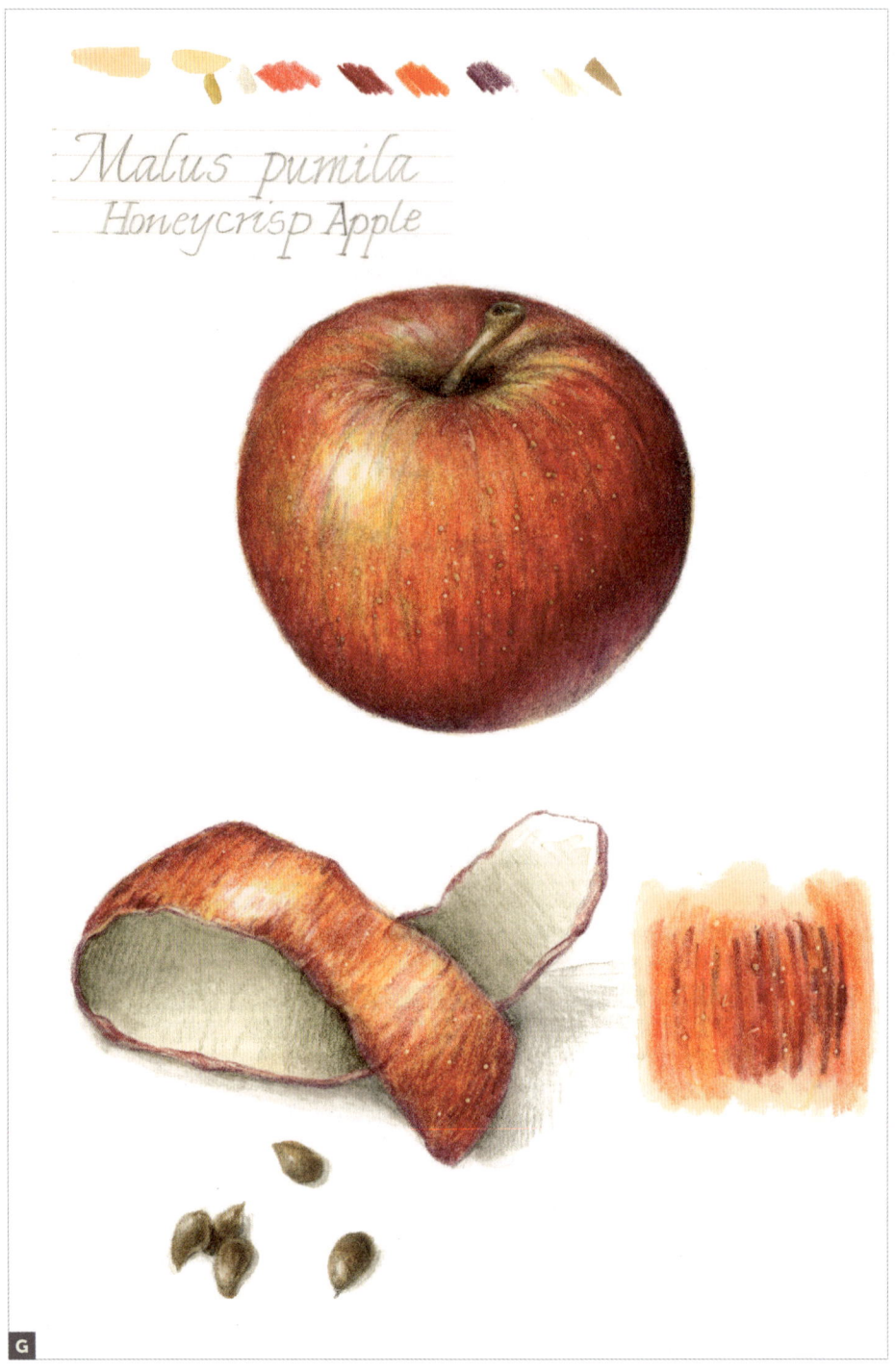

Malus pumila
Honeycrisp Apple

G

Drawing the Spiral Pattern on a Pine or Spruce Cone

Trees that bear cones are called conifers. They produce needles and cones instead of leaves and flowers. I often collect cones when I walk outside. Usually in the cold months, the cones are closed. After a few days inside, however, they warm up, dry out, and open up. The seeds are inside. If you want to keep your cone closed for an easier drawing, soak it in water and it will close up again. I discovered this on my own with a bit of research, and learned the reasons why a cone opens and closes. I could tell you here, but I think it's better for you to discover this independently. Hypothesize your own answers, and then do some research and find out why! Nature always has a reason for doing things in a particular way, and it's exciting to learn about a plant's function. Form and function go hand in hand in nature. Form usually maximizes use of space to tightly pack as many seeds as possible into a small area.

Studying spiral patterns will help you draw them more accurately and with ease. What at first might seem to be an overwhelming job becomes a simple unlocking of the underlying pattern. This pattern can first be copied as a simple outline and then details of form can be added. One way to clearly see the patterns in nature is to take a photograph of your subject. Place a piece of tracing paper over the photograph and draw the lines created by the individual scales or sections of the form. Once you start paying attention to the patterns in nature, you'll see them everywhere. This lesson focuses on spiral patterns formed by the scales of female spruce cones, but you can repeat this lesson as much as you want with a variety of subjects, such as an artichoke, a spiral aloe, or acorns. Below is an additional drawing of an open pine cone of the Himalayan pine tree along with needles and seeds.

SUBJECT

Pine or spruce cone

ADDITIONAL MATERIALS

- Dark Sepia colored pencil #175
- Watercolor brush #2
- Watercolor pencils and colored pencils to match the local color of your cone
- Ivory colored pencil #103
- Verithin pencils
- Magnifying glass or other magnifier

COLORED PENCILS

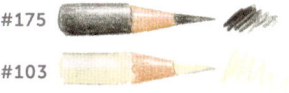

#175

#103

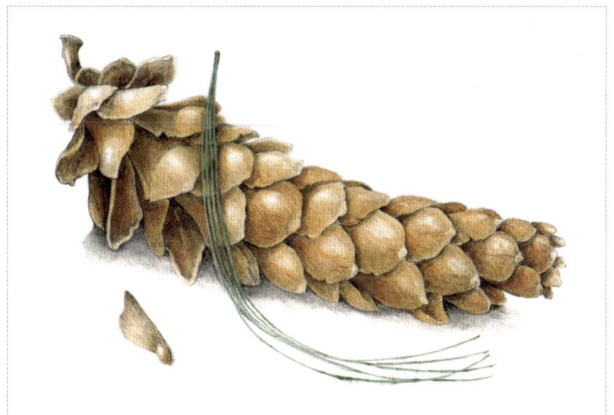

CONTINUED

1. First analyze and draw the pattern you see on the cone. Lightly draw the angles formed by the rows of scales that travel in two directions. You don't have to copy the pattern exactly once you've measured the angles and the width between scales accurately. In my cone, the angle is close to, but perhaps a bit more than, 45 degrees in one direction. In the other direction, the angle is less than 45 degrees. No need to be exact with this once you understand the idea, but this concept will make your drawings look more natural. **(A)**

2. Try drawing an enlarged area of the scales to understand their shape. **(B)**

3. With a Dark Sepia colored pencil, redraw the cone lightly, showing the overlaps of each scale with a bit of tone. Think about grisaille toning the whole cone with your idealized light source of a cylinder. While grisaille toning each scale individually, also try to render the scales on the shadow side darker than those that receive more light. Don't forget to leave some good highlights. **(C)**

4. Using a #2 brush, build up layers of color with a watercolor wash and then layer colored pencils. Continually remind yourself to maintain an overall tonal variation based on shadows and highlighted areas in each scale as well as on the overall cone itself. This is a good time to work slowly and not rush the drawing. Build layers of color slowly, keeping good contrast between individual scales; burnish with an Ivory colored pencil and add fine details with Verithin pencils. **(D)**

5. Add in a branch with scales, needles, and enlarged details to further identify your particular species.

6. See if you can pull apart some scales (this won't be easy, as the structure is amazingly strong) to draw them individually. Find some of the seeds inside to look at under a magnifying glass, and draw them enlarged as well. **(E)**

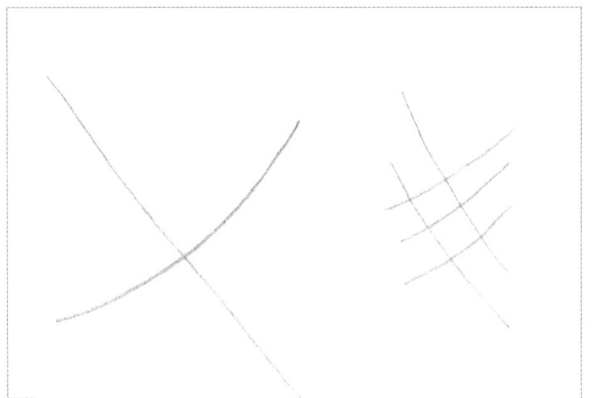
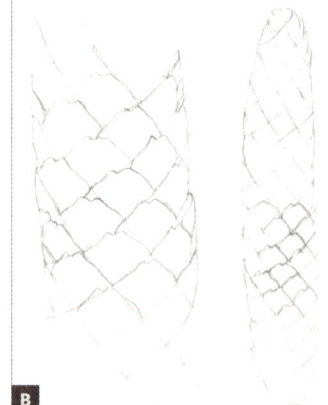

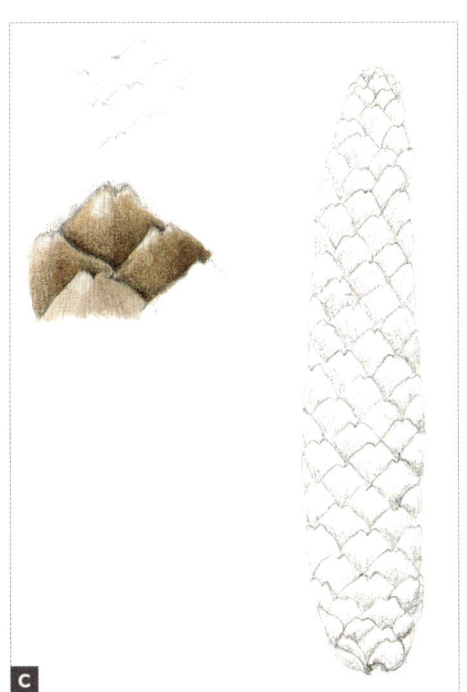

C

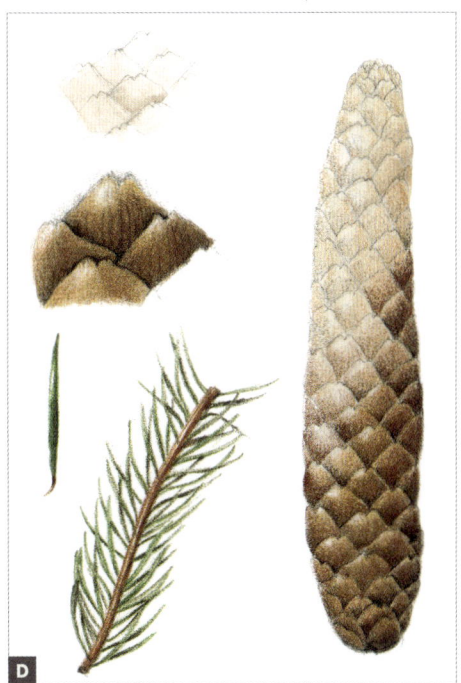

D

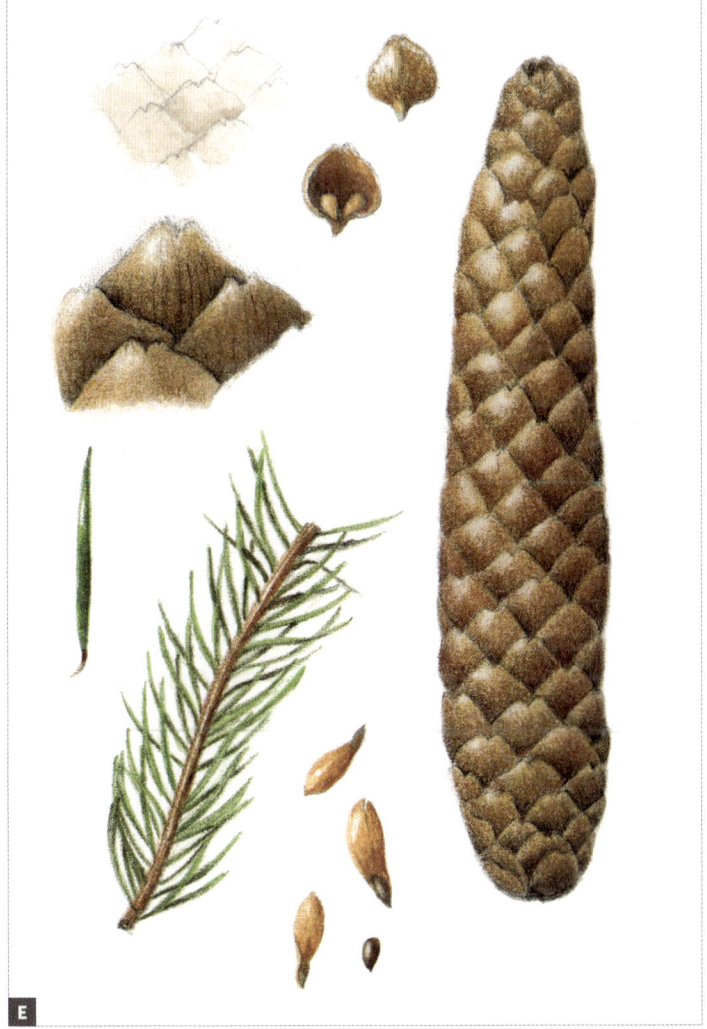

E

Drawing the Pattern on a Strawberry

SUBJECT

Strawberry

ADDITIONAL MATERIALS

- Magnifying glass or other magnifier
- Watercolor brush #2
- Colored pencils and watercolor pencils to match the local color of your strawberry
- Ivory colored pencil #103
- Verithin pencils

COLORED PENCILS

#103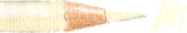

Though we call them "berries," the technical botanical term for strawberries is *aggregate fruit*. In this drawing of a strawberry, I drew the berry enlarged two or three times. This helped me capture not only its spiral pattern but also the fact that the strawberry's tiny seeds are on the outside. I mathematically enlarged by first measuring accurately and then multiplied by 3 for my strawberry so I could keep the proportions accurate.

1. Measure and draw, with a graphite pencil, a strawberry, life-size or enlarged. Draw rough lines to indicate the spiral pattern created by the seeds. (This is the same pattern as in the previous lesson on the spruce cone.) **(A)**

2. Study the surface of your strawberry under a magnifying glass. To practice the pattern, show each seed slightly indented into the strawberry's flesh with a shadow. Draw each seed on your subject within the pattern lines on a small practice sample. Add color to each seed. **(B)**

3. Using a #2 brush, add a layer of watercolor around or over the seeds, but leave the area around the highlight on the strawberry empty for now. **(C)**

4. Continue adding layers of color to your strawberry, remembering to keep the shadow side of the berry darker. Burnish with an Ivory colored pencil, refine the highlighted area, and refine edges and seed details with Verithin pencils. **(D)**

5. Once your berry is finished, complete the composition with the addition of a flower, a cut fruit, and a grouping of strawberry leaves. **(E)**

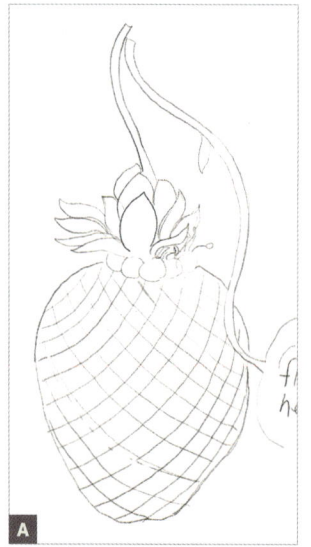

A

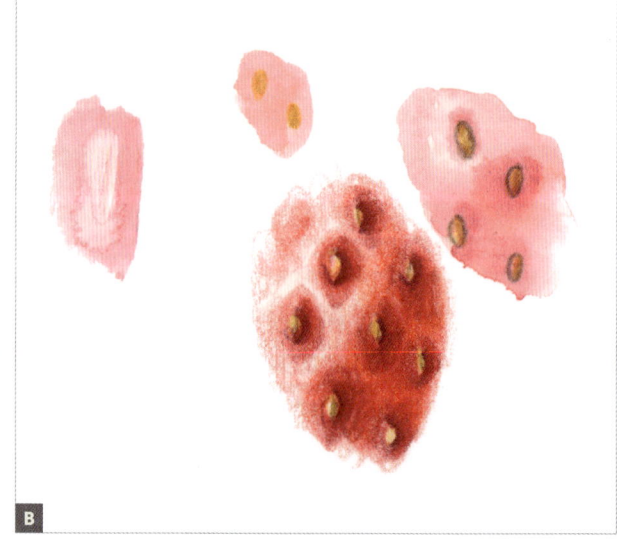

B

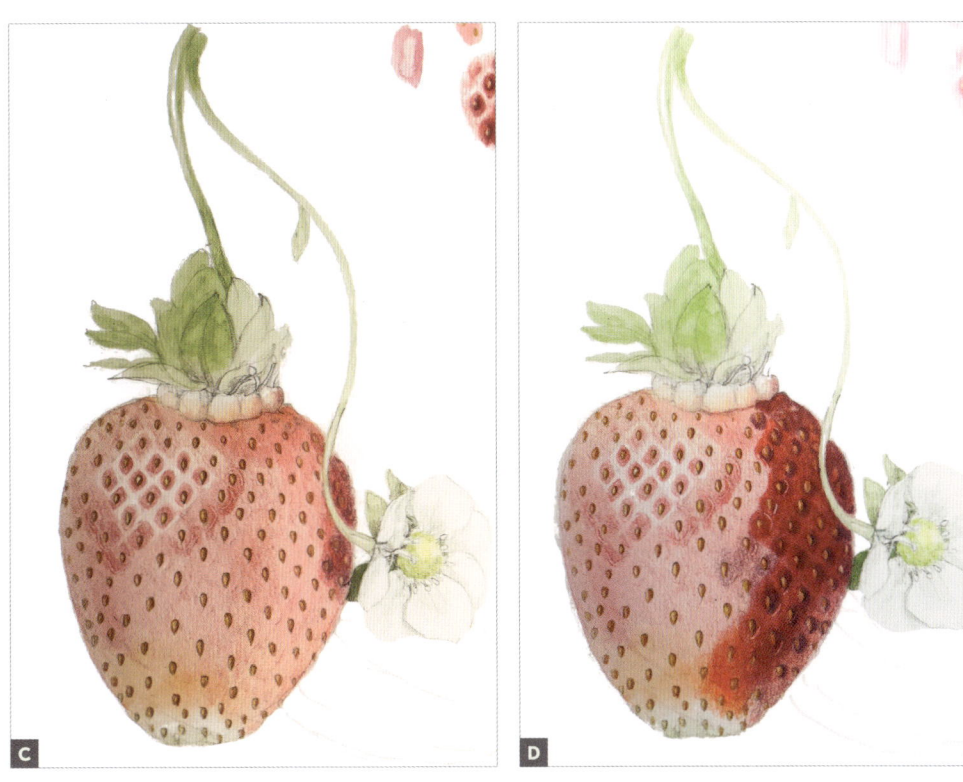

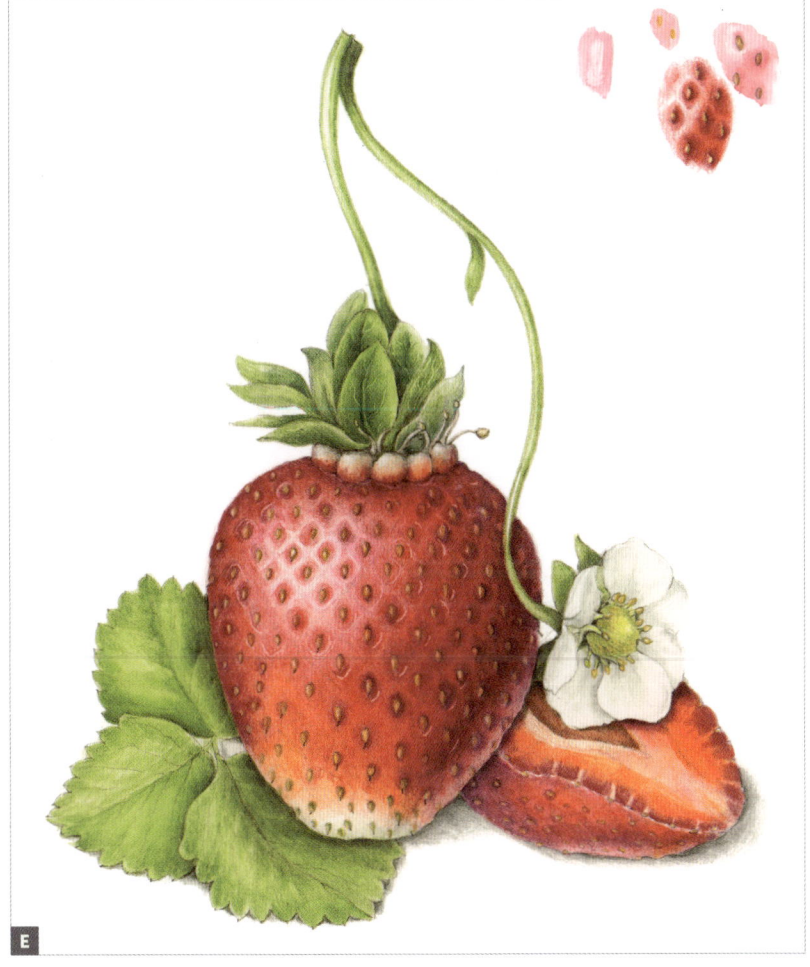

Drawing the Pattern of Spikes on a Sweet Chestnut

SUBJECT

Seedpod with spikes, such as sweet chestnut, datura, *Bixa orellana*, or sweet gum

ADDITIONAL MATERIALS

- Embossing tools (small to medium)
- Dark Sepia colored pencil #175
- Magnifying glass or other magnifier
- Colored pencils and watercolor pencils to match the local color of your subject
- Verithin pencils

COLORED PENCILS

#175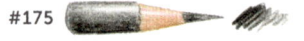

Seed-bearing fruits sometimes have spikes, which are usually wider at the base and then taper to a very fine, sharp point. This thorny surface is a means of protecting the seeds from predators. Drawing so many fine lines is difficult, so use embossing tools to preserve a sharp line. Nature is full of unusual interesting subjects, not necessarily all beautiful. I encourage you to explore the off-putting (in this subject, literally) and mysterious side of nature, along with the more typical pretty side.

1. With embossing tools, practice drawing some individual spikes and try to get a variety in thickness. Tone on top of them using a Dark Sepia colored pencil so that you can see the results. **(A)**

2. Study your seedpod under a magnifying glass. Draw the spikes enlarged to show and help you understand their structure. Follow all the usual steps to add color and details. Verithin pencils are great for these kinds of fine details. **(B)**

3. With Dark Sepia, draw the nut and its details lightly. Emboss some spikes and hairs and then apply more definition with colored pencils. **(C) (D)**

4. Add more color to the nut with colored pencils and darken behind the spikes in front. **(D)**

5. Continue to develop the spikes with drawing and embossing of the layers of spikes. Add a leaf to create contrast with the spikes in front of the leaf. **(E) (F)**

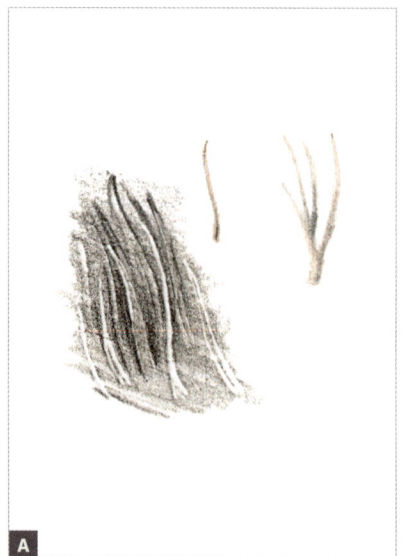

A

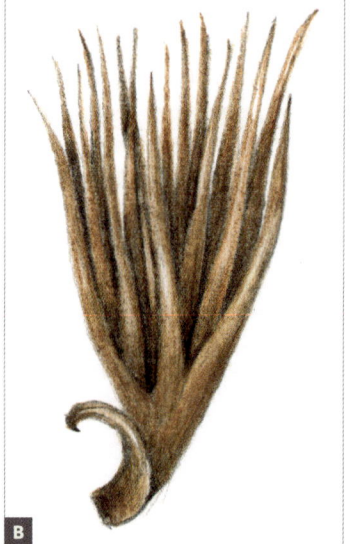

B

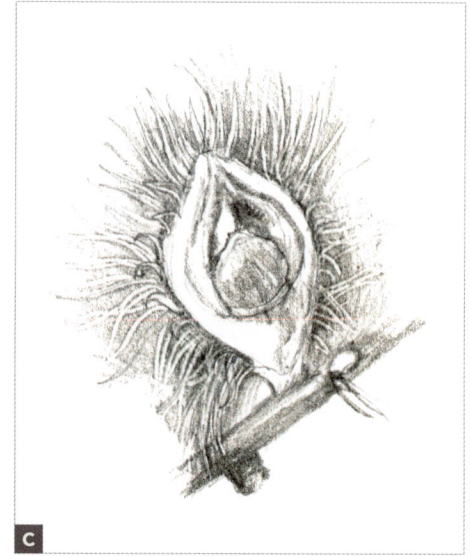

C

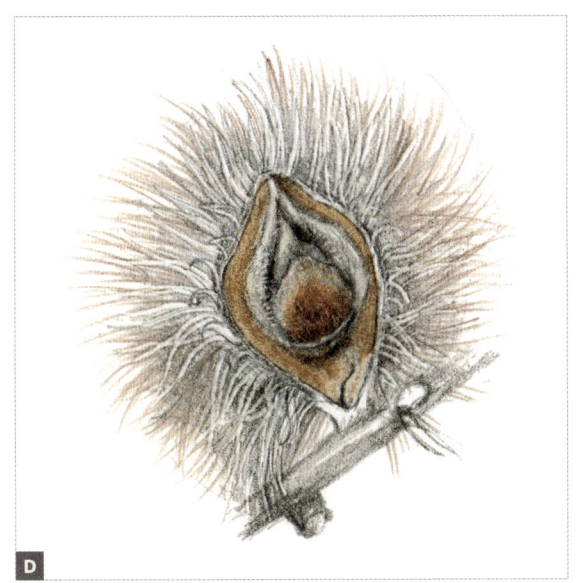

D

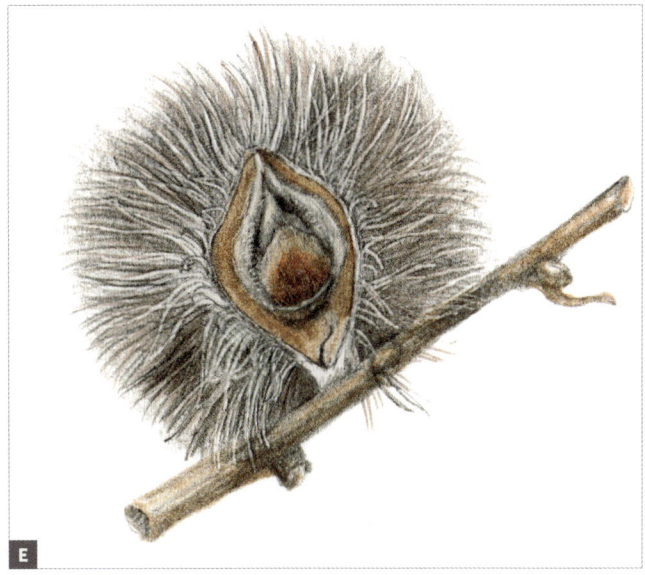

E

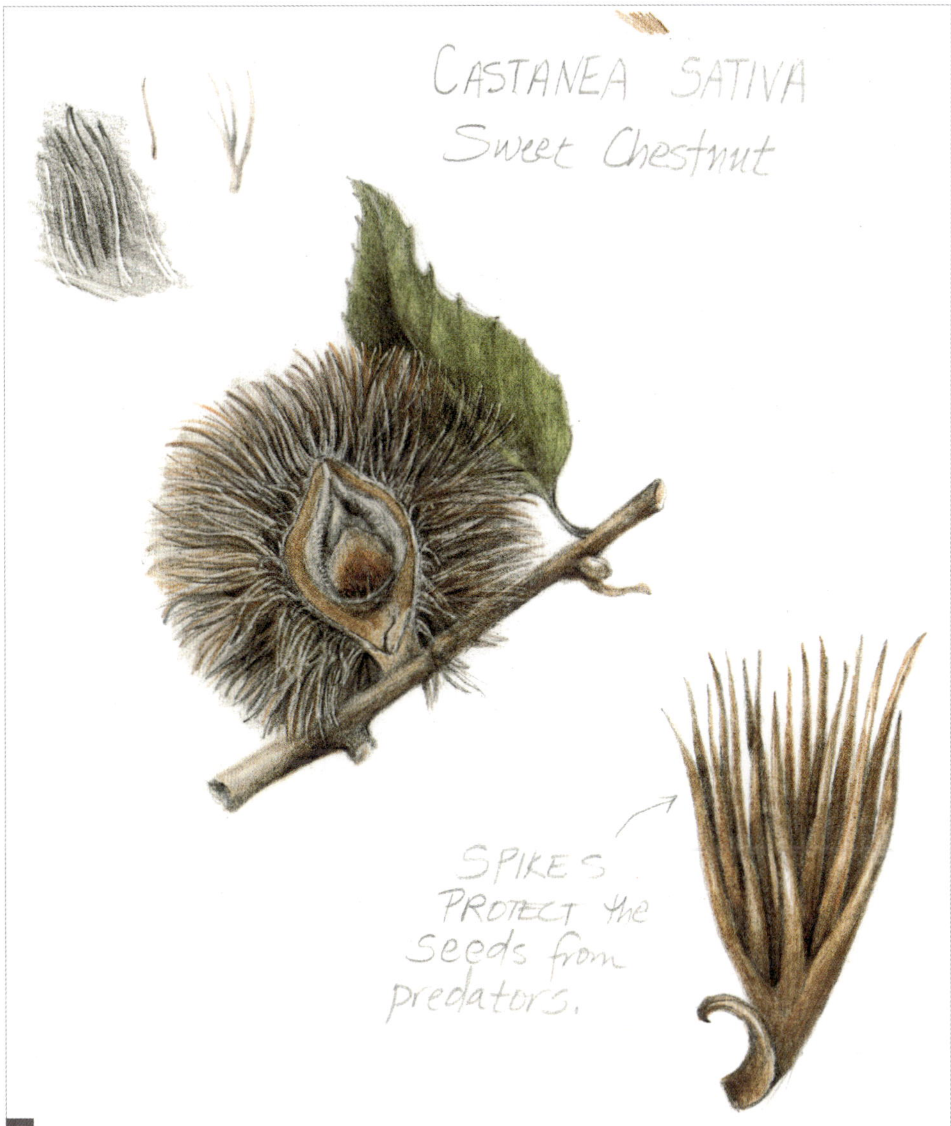

CASTANEA SATIVA
Sweet Chestnut

SPIKES
PROTECT the
seeds from
predators.

F

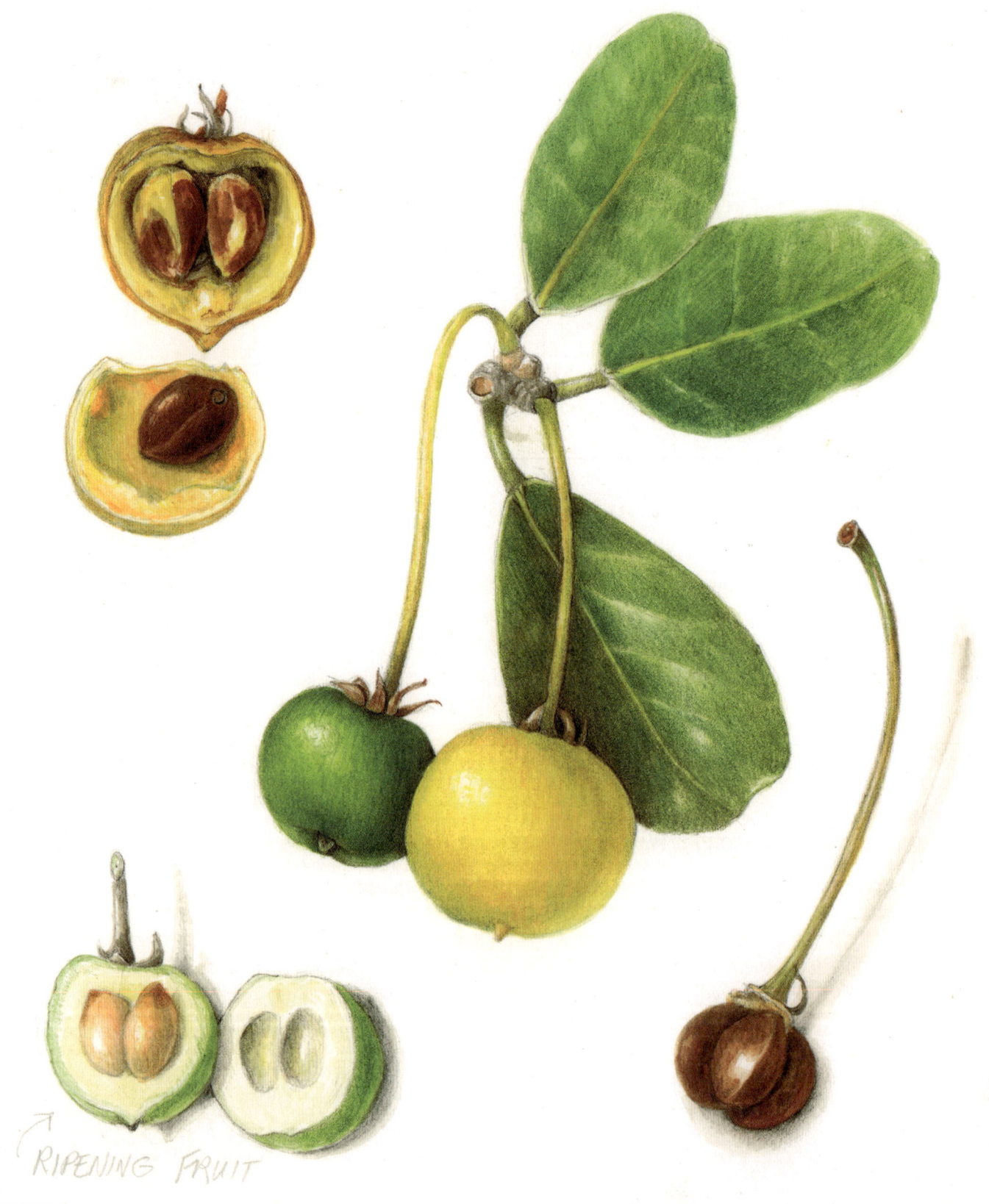

Annona glabra
Monkey Apple

RIPENING FRUIT

CHAPTER NINE

BITE-SIZE BOTANICAL COMPOSITIONS

Now that I've presented step-by-step lessons on all the components of a plant, let's talk about planning pages or compositions. This starts with creating a compelling focal point. Simply put, this is the spot on which you want the viewer to focus. I let nature be my guide in planning a page. I look at each plant and try to tell a story about it, even if it is only about a small part of the plant. As plants are masters of attraction, there are built-in focal points everywhere! Nature uses color and form to attract pollinators—and that includes us.

Go outside and find a plant that attracts you. Consider what you find most interesting about it. What do you want to stare at? Decide on the most compelling part of the plant and make that your focus. Often this can be an identifying characteristic, such as a flower, unusual seedpod, or leaf arrangement. Any of these elements will usually work as a focal point. I usually start by drawing a focal point. Once I've established this, I examine and sometimes take apart my subject to see what else there is to explore. I might cut a fruit in half, crack open a nut, or pull apart a flower. I take delight in my discoveries and invite you to do the same. If you have a personal story about a plant, you can tell that, or do some research to learn some interesting facts about the plant. This is a great way to make decisions on what you want to show in your drawing. In today's world, it takes only seconds to look up a plant and read about it.

Many of the drawings I included in this book are done on 5 by 7-inch paper. As I've previously mentioned, I love this size for small compositions. It isn't so big that I have to commit a lot of time, and I know I can manage small drawings in my busy schedule. They fit the bill perfectly because they allow me to draw almost every day. Of course, some subjects are bigger, so not appropriate for this small paper, but so many easily fit here. This is one way to keep going with your own practice. Sometimes you will work on larger drawings and compositions, but often this small size is perfect for drawing on a regular or daily basis.

TIPS FOR CREATING A COMPELLING COMPOSITION

Here are some of my favorite tips on making a great page composition. I use all or some of them, depending on what I want to accomplish. Use them as a checklist to help you make decisions as you go.

- Start with an interesting focal point.
- Render more detail in the foreground.
- Render less detail in supporting elements and those farther back in space.
- Use more contrast and saturation of color in front, making sure to have your darkest darks and your lightest highlight in the foreground, with fewer finishing details in back.
- Use warmer and brighter colors in front, as these colors appear to advance toward the viewer.
- Use cooler and grayer colors behind, as these colors seem to recede.
- To tell a complete story of a plant, show most of its parts, including leaves and how they are connected to the stem, flowers, fruit or seed capsule, roots, and sometimes a dissection or enlarged parts.

Include text with information about the plant, including common name, scientific name, location where you collected the subject, date, and notes containing interesting facts and questions about the plant.

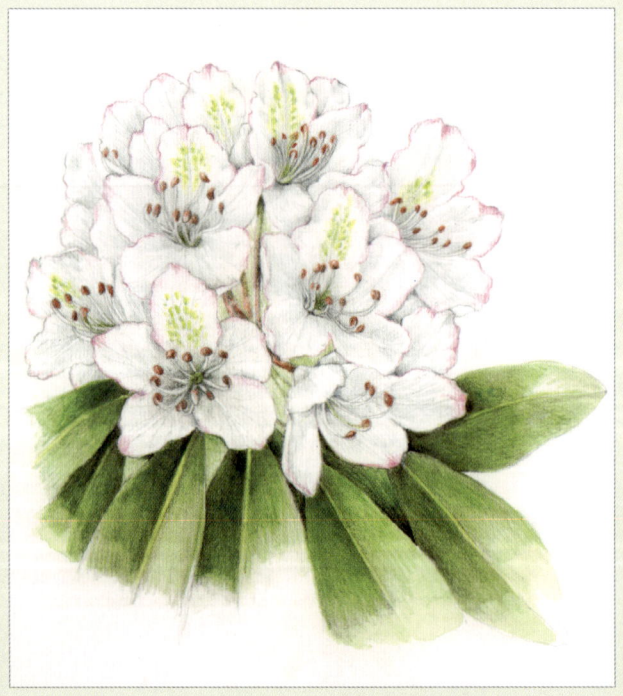

Plant Process Pages

I think I love my process pages best because they do two things at once—they document my exploration of a plant and they show my drawing process. I used to call these pages sketchbook pages or study pages, but now I think "plant process page" (PPP) sums it up. If you create your own PPPs, you'll have a safe space in which to explore drawing without worry and judgment. Your page is about studying a plant and practicing drawing techniques, therefore it's not a "finished piece of art" and not required to live up to any expectations. Yet I promise you it will. Unfinished drawings are so mysterious, they show enough, but not everything, allowing your imagination to fill in the rest. They also provide a great way to keep drawing yet not have to color in every element on a page. It allows for the viewer to see what interests you most, and not focus on a jumble of leaves, for example. Your notes, questions, and discoveries can be recorded on the page, along with the color pencils you used or a new technique you were practicing.

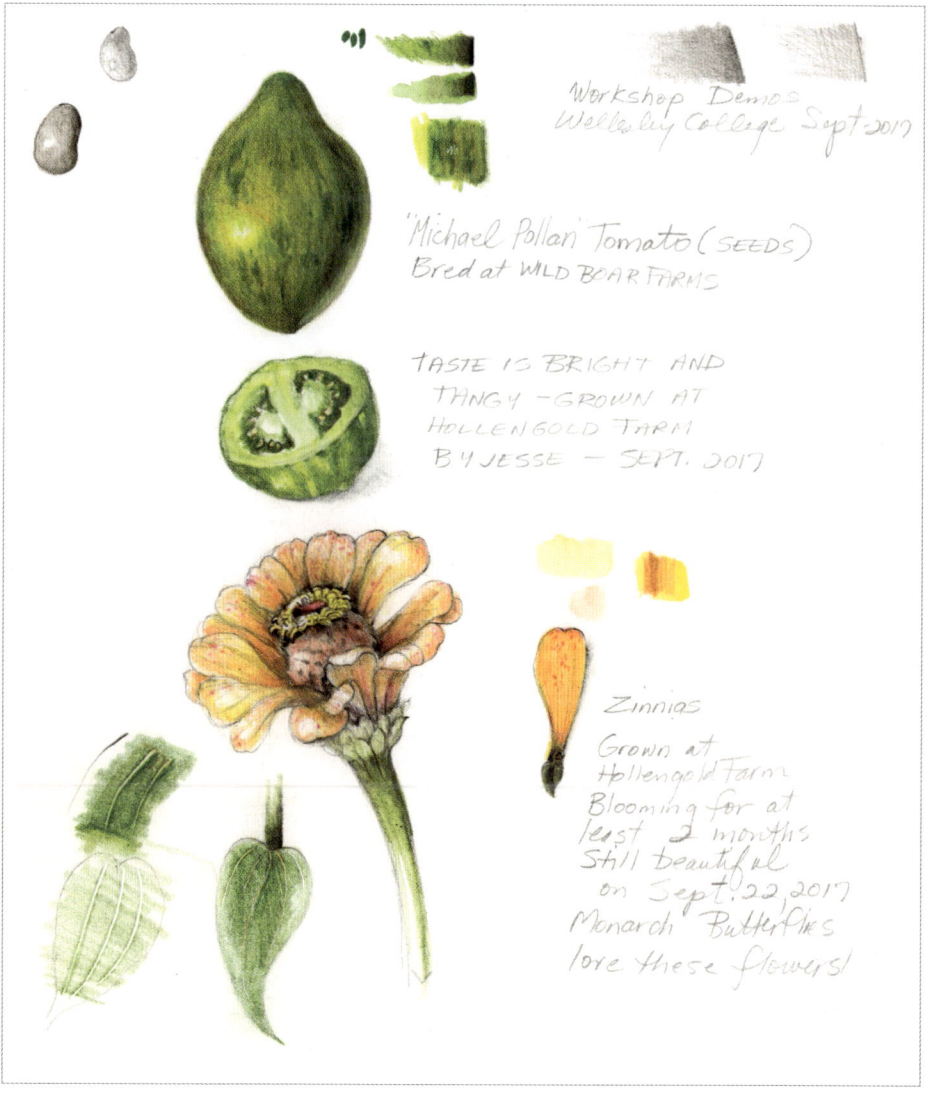

Workshop Demos
Wellesley College Sept 2017

'Michael Pollan' Tomato (SEEDS)
Bred at WILD BOAR FARMS

TASTE IS BRIGHT AND
TANGY —GROWN AT
HOLLENGOLD FARM
BY JESSE — SEPT. 2017

Zinnias
Grown at
Hollengold Farm
Blooming for at
least 2 months
still beautiful
on Sept. 22, 2017
Monarch Butterflies
love these flowers!

Watermelon Radish

ADDITIONAL MATERIALS

- Watercolor brush #2
- Watercolor pencils and colored pencils to match the local color of your radish
- Embossing tool (small)
- Wax paper
- Black Verithin pencil
- Cool Grey Verithin 70% pencil

Once you're comfortable with botanical illustration and you understand how to use a specific light source, you can begin a drawing with a layer of watercolor. Plan your page by putting your subject on the paper to find an interesting composition and to make sure it will fit nicely on the page. Decide if you'll be adding overlapping elements and plan where to put them.

1. With a graphite pencil, lightly draw a complete watermelon radish composition, including leaves, root, and cut sections. Using a #2 brush, add a watercolor layer of all appropriate colors, leaving highlights that are the color of the paper where needed. **(A)**

2. When this layer is dry, decide where overlapping roots and texture will be, and use an embossing tool to draw some root and vein details on the leaves. **(B)**

3. Layer colored pencils on top to create three-dimensional light and shadow areas in appropriate colors, and, of course, burnish as needed, especially around the highlight. **(C)**

4. Use colored pencils over embossed areas to emphasize the embossing as needed. **(D)**

5. Render the leaves with green, and then use the embossing tool with wax paper to create lighter veins. Accentuate darks and saturated color with colored pencils, and use the Black and Cool Grey 70% Verithin pencils to sharpen edges and details. **(E)**

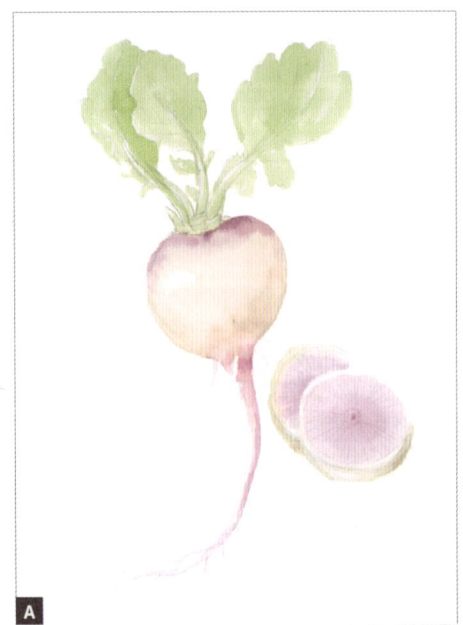

A

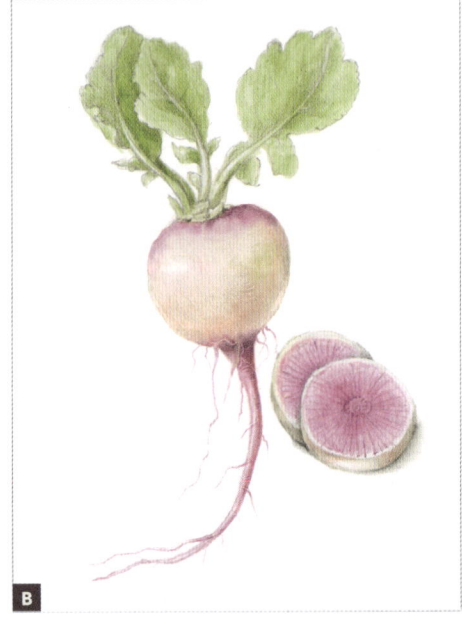

B

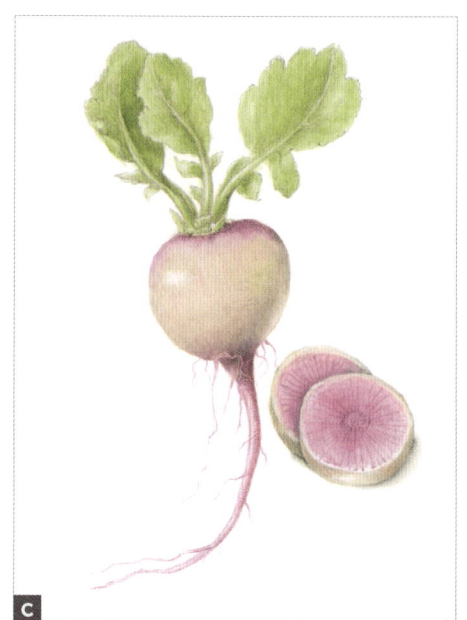

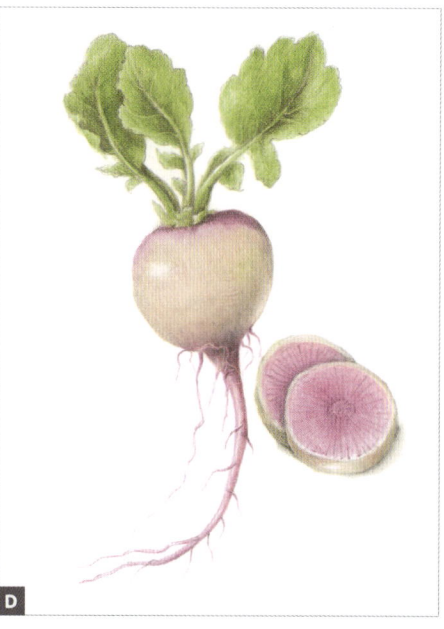

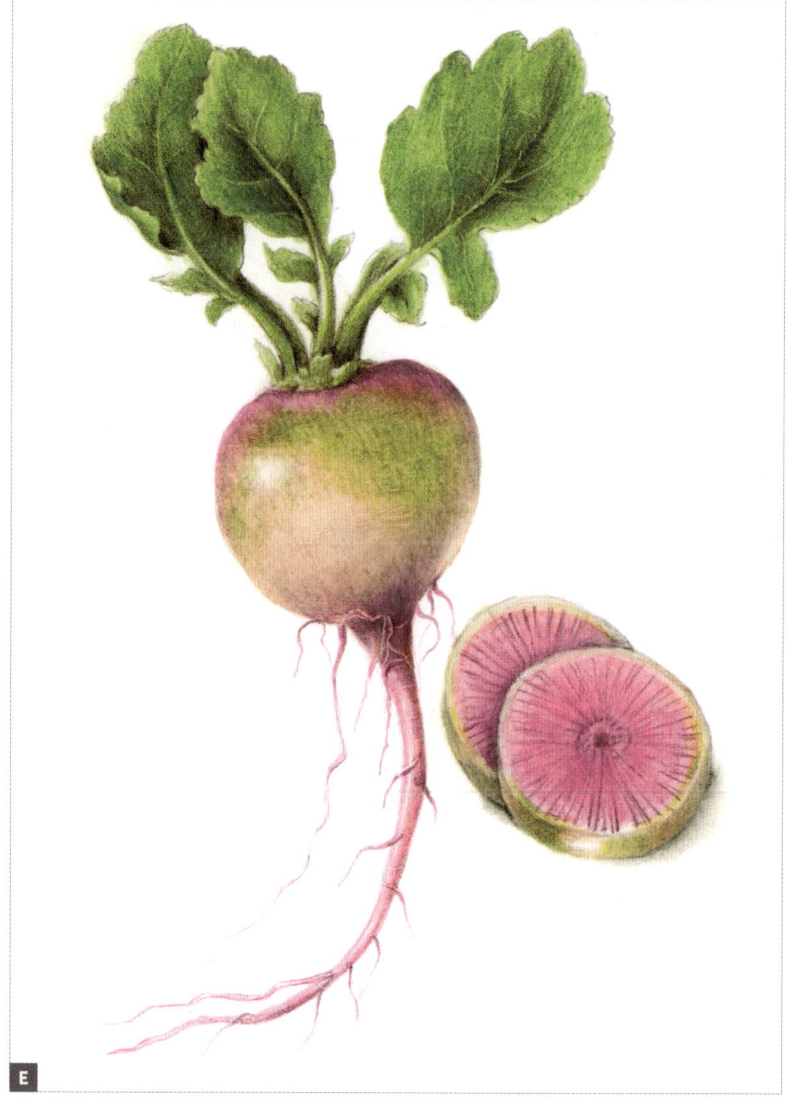

Hydrangea (Inflorescence Flower)

SUBJECT

Hydrangea

ADDITIONAL MATERIALS

- Colored pencils and watercolor pencils to match your subject's local color

- Magnifying glass or other magnifier

An inflorescence is a cluster of flowers on a stem, such as seen in hydrangeas, lilacs, or azaleas. It's composed of many flowers with smaller stems, each connecting to the larger stem. On a hydrangea, the flowers are actually the tiny round areas in the center of what look like petals, but technically they are the bracts (modified leaves).

Fun fact about the hydrangea: the name derives from the Greek words *hydro* (water) and *angeion* (vessel). This refers to the seed capsule, which is very tiny, but nevertheless resembles a pitcher. The color combinations on hydrangea vary greatly, and this particular one has shades of blues and greens, which make for a vibrant subject to draw. The key to creating an interesting composition of an inflorescence flower is to remember that a compelling focal point will bring the viewer into the page. Using precise perspective drawing, strong contrast, and grisaille toning on the whole form and not just each individual flower allows for a dynamic composition. We can use a simple round form as a light-source model for the entire inflorescence. This means that flowers on the shadow side will be darker than those on the top and highlight area. Notice how I also include good overlaps of dark shadows where the flower head meets the leaves. This makes the viewer want to come inside my flower. In the drawing of a rhododendron on page 154, I also used this same technique to bring you into the drawing.

1. Study the entire cluster, and draw thumbnail sketches of the light and shadows on the whole form. **(A) (B)**

2. Notice that the cluster has distinct light and shadow areas, not only in each flower but in the form as a whole. **(C)**

3. Notice how one flower shape repeats many times from slightly different perspectives. Remove a few flowers and practice drawing several views. Position your ellipses in different directions for this. Also draw some flowers more open than others for variety. **(D) (E)**

4. Use a magnifying glass to study the centers and reproductive parts of the flowers.

5. Practice color variation on individual flowers and then vary your local colors for the entire cluster, using a range of colors and strong darks especially on the shadow side. Making use of overlapping techniques to separate the individual flowers is important here.

6. Add in leaves following all the usual steps, especially creating good dark shadows so that the flowers feel like they're sitting on top of the leaves. **(F)**

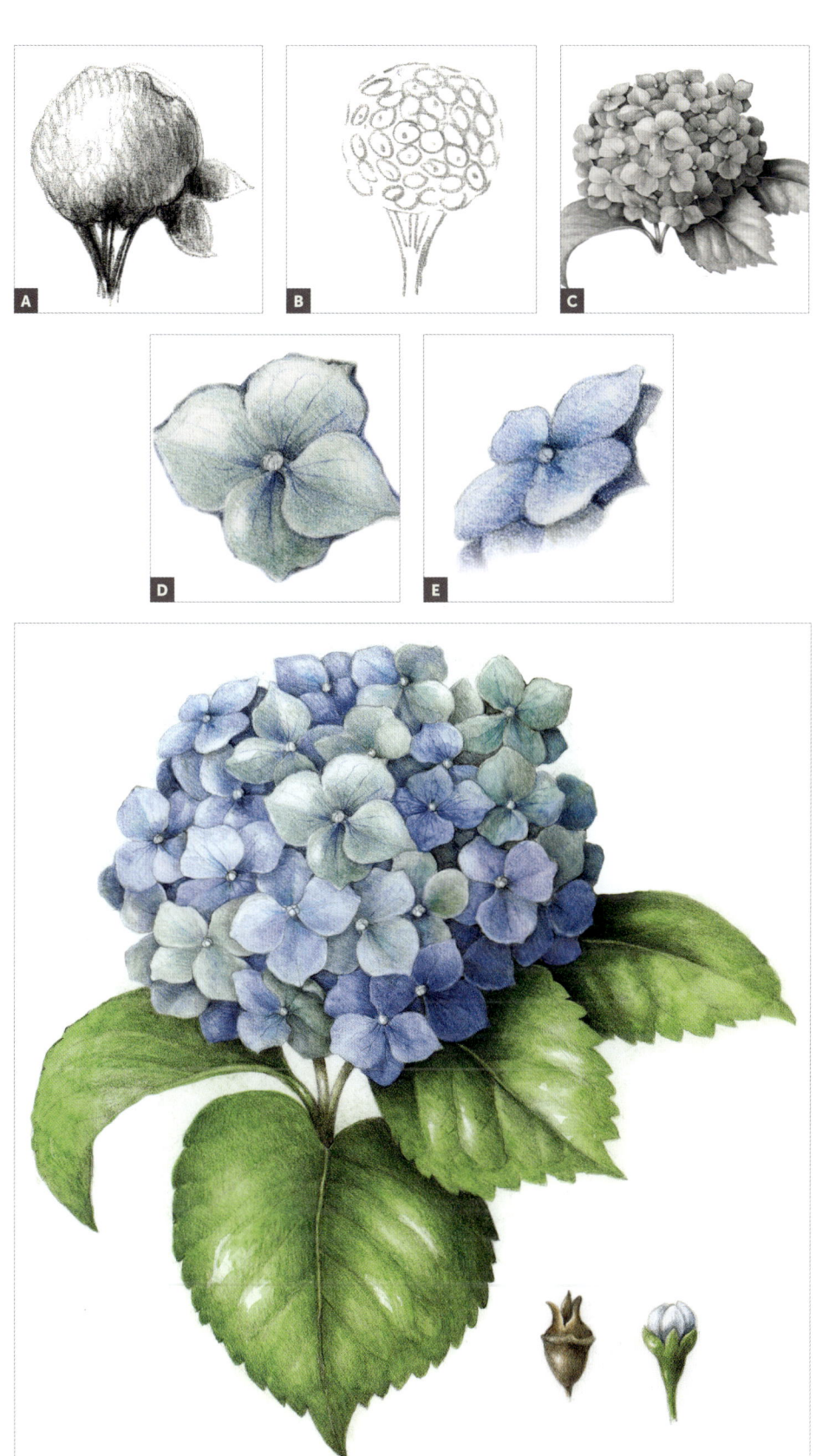

Page Composition Study

SUBJECT

Fallen nut or other element found on or around a fruiting tree

ADDITIONAL MATERIALS

- Dark Sepia colored pencil #175
- Colored pencils and watercolor pencils to match your subject's local color
- Watercolor brush #2
- Embossing tool (optional)
- Ivory colored pencil #103
- Verithin pencils

COLORED PENCILS

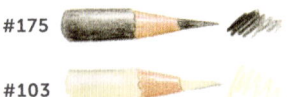

#175

#103

A detailed composition can tell a story about a plant. In these next bite-size botanicals, I have gathered elements found on or around a fruiting tree. I plan the composition with the elements placed on the page, starting with the main event and adding other elements to fill out the space. In some cases, I leave space on the page to revisit the tree when it might be flowering so I can add it to my composition. Fallen nuts are excellent subjects. They keep their color and shape for a long time and travel well, so they are good companions for drawing on the go. These next two compositions follow the same steps so use them as a template over and over again.

1. Scavenge and collect elements from a particular tree to draw.

2. Write down the name and location of the tree so that you can revisit it later. Arrange the elements in a pleasing composition, planning a focal point.

3. Measure and draw your subject life-size with a graphite pencil on good paper.

4. Put on your head light, and add grisaille toning with a Dark Sepia colored pencil.

5. Choose local colors and build layers with colored pencil and watercolor washes, using a #2 brush for the washes and emphasizing your focal point with good contrast and dark shadows.

6. Add more layers of colored pencil and watercolor if needed. Use some subtle embossing if appropriate.

7. Burnish your subject with an Ivory colored pencil, and add details to the highlights.

8. Add details with sharp colored pencils and Verithin pencils for edges and cast shadows.

OHIO BUCKEYE HORSE CHESTNUT (*AESCULUS GLABRA*)

I found an Ohio Buckeye tree in New York's Central Park in early October. The fruits were really interesting and soon started to split open, revealing the nut inside. The leaves have five leaflets and start to turn colors from brown to yellow to green. I have a little room left in the composition, so perhaps next spring I'll add a flower if I remember exactly where the tree is!

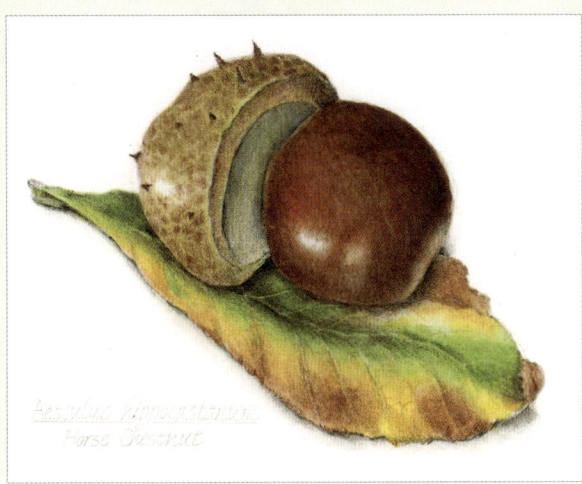

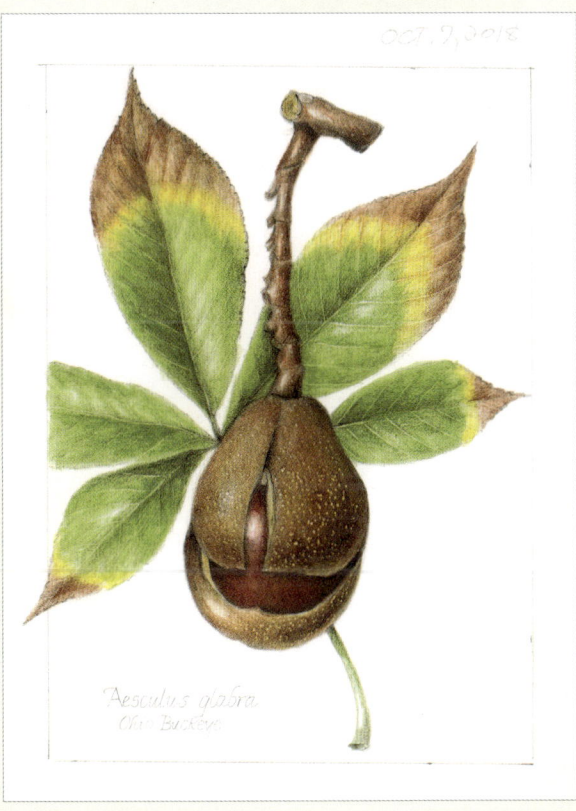

BLACK WALNUT (*JUGLANS NIGRA*)

In January, there are still many black walnut fruits on the ground in the Hudson Valley where I live, and some are in various stages of opening and drying. I noticed that some of the nutshells seem to have two holes hollowed out in the exact same spot. I'm pretty sure this is the work of squirrels. They know precisely where to bore the hole to get to the sweet nut meat inside!

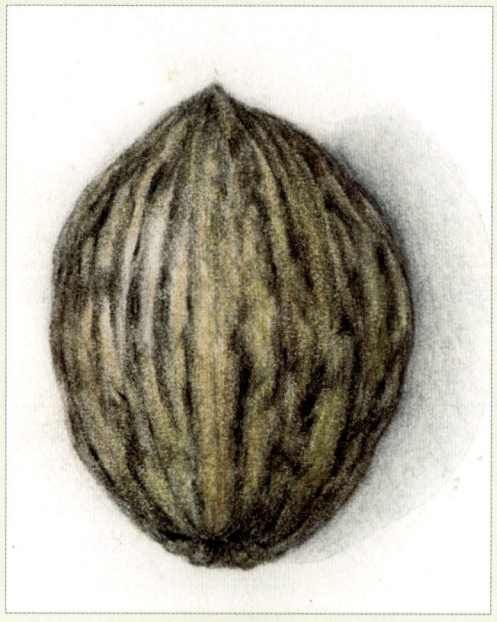

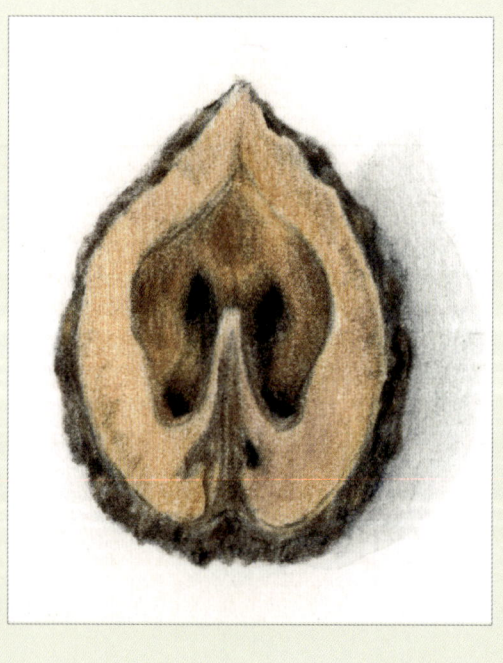

JANUARY 1, 2017

Outer husk

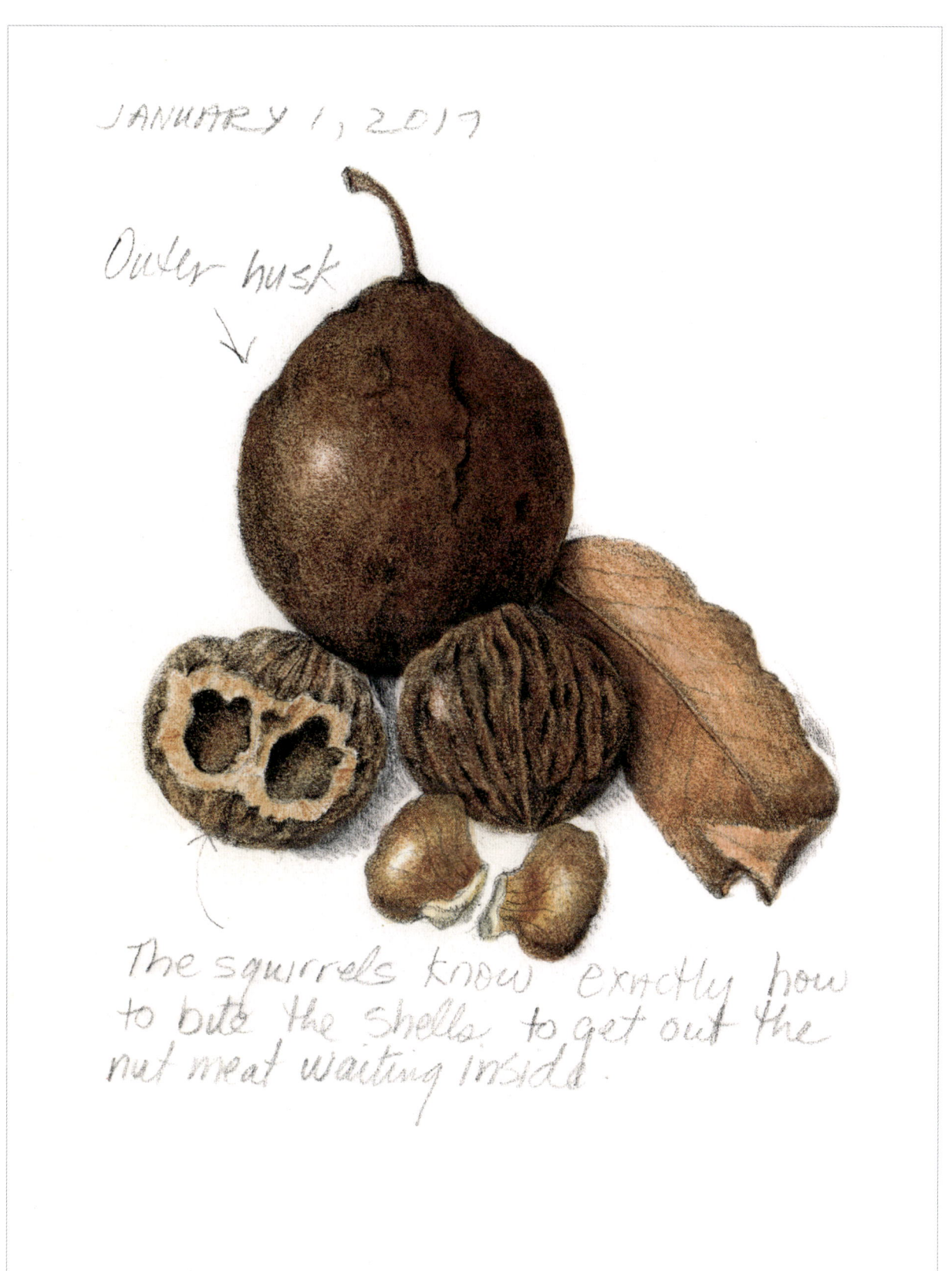

The squirrels know exactly how
to bite the shells to get out the
nut meat waiting inside.

Avocado (*Persea americana*)

SUBJECT

Avocado (preferably with branch and leaves intact)

I love edible subjects, as they are near and dear to my heart and stomach. Avocados are a favorite fruit, and I look forward to picking them directly from the trees when I'm in a tropical environment. It's amazing how prolifically they grow. This avocado is from a friend's garden in Hawaii. I wanted to capture the way the fruit hangs on the tree with some leaves. My goal was to make the fruit really shine and show the bumpy surface. I also wanted to focus on a cut-open fruit. When I opened this fruit, I noticed that the nut made a heart-shape indentation in the fruit, so I made sure to include it in the drawing, but I kept it subtle. I also wanted a focus on the large, shiny nut and the contrast of the rich, warm brown color against the yellow flesh of the fruit.

1. Set up a pleasing composition, and draw your avocado life-size with a graphite pencil on good paper.

2. Develop the drawing of the fruit, leaves, and stem with the usual steps we've practiced.

3. When finished, cut open the fruit to add to your composition. When rendering these parts, be sure to pay attention to the overlapping areas of the fruit in front to create good contrast and an exciting focal point. It's helpful when planning a composition to be aware of the direction of the elements and avoid angles that might pull the viewer's eye off the page. Notice how I have positioned my avocado pieces to bring the viewer right back to the main hanging fruit.

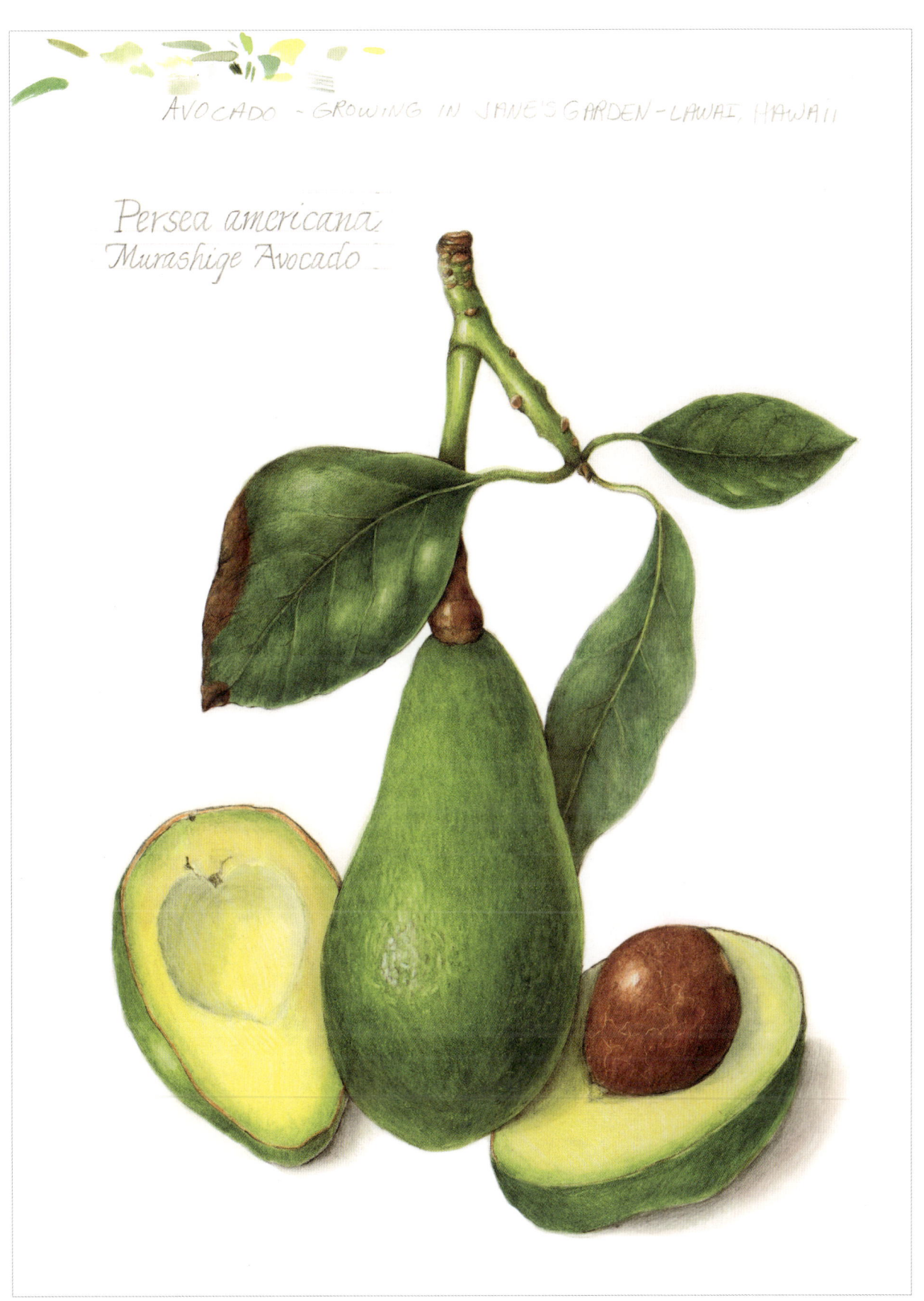

AVOCADO - GROWING IN JANE'S GARDEN - LAWAI, HAWAII

Persea americana
Murashige Avocado

Morel Process Page

My first foray into foraging mushrooms occurred a few years ago with morels in the spring. It is truly a magical experience. It takes a while to develop your mushroom eyes, but once you do, you may notice them everywhere. Many times I've had this exact experience: I will search methodically for about an hour and find nothing. Then, when I retrace my steps, I'll start to find morels. It feels as if the mushrooms were hiding and waited until I made myself worthy by searching patiently. As a reward, they suddenly became visible. With this process page, I tell my foraging story and include my favorite recipe.

1. Forage patiently with an experienced forager, looking for edible morels.

2. When you find them, go immediately to your drawing board and draw them. They will stay fresh when refrigerated for a few days while you complete your drawing.

3. Only after you complete your drawing can you cook and eat them.

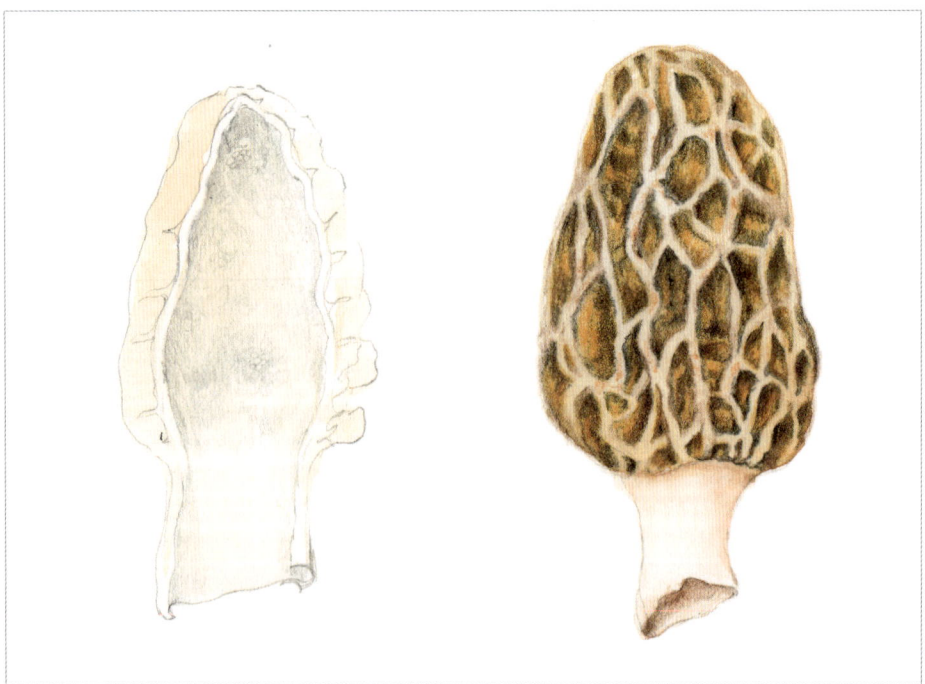

May 2, '16
Black Morel

Morchella - MORELS

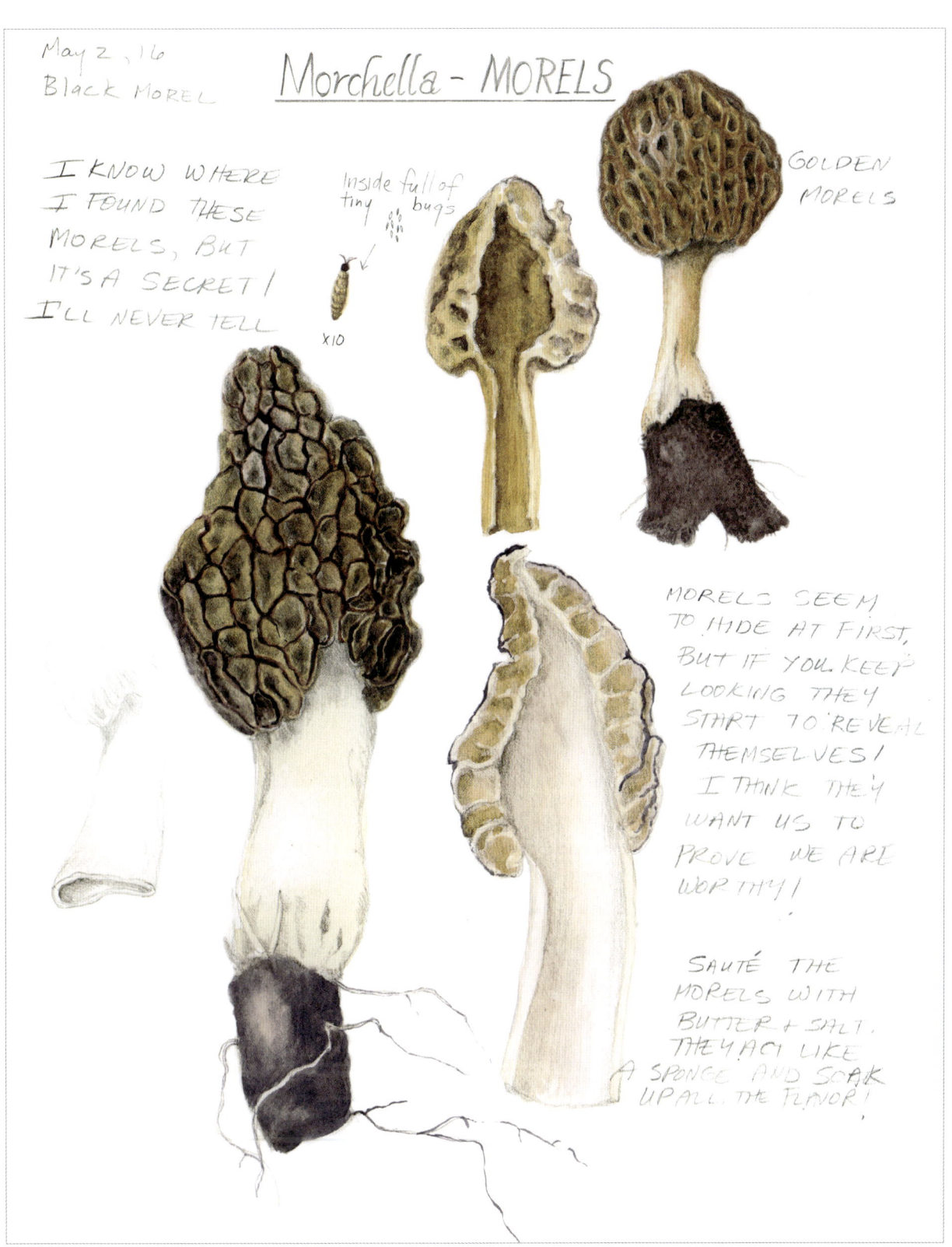

I KNOW WHERE
I FOUND THESE
MORELS, BUT
IT'S A SECRET!
I'LL NEVER TELL

Inside full of
tiny bugs

x10

GOLDEN
MORELS

MORELS SEEM
TO HIDE AT FIRST,
BUT IF YOU KEEP
LOOKING THEY
START TO REVEAL
THEMSELVES!
I THINK THEY
WANT US TO
PROVE WE ARE
WORTHY!

SAUTÉ THE
MORELS WITH
BUTTER + SALT.
THEY ACT LIKE
A SPONGE AND SOAK
UP ALL THE FLAVOR!

SOURSOP COMPLEX COMPOSITION

This PPP of a soursop is part of an exhibition called *Out of the Woods: Celebrating Trees in Public Gardens,* cosponsored by the New York Botanical Garden and the American Society of Botanical Artists. This juried exhibition has traveled to botanical gardens across the United States and shows much of the highest caliber of botanical art from artists all over the world. I had so much to tell about this tree that I decided to use an informal format that included notes and color dabs. I want to allow the viewer to look and learn, and thus share some of the journey I went on with this plant. Plant process pages allow for this.

I was surprised when they chose my informal process page to be part of this exhibition but was happy to see it there. Though I admire and am in awe of the painstaking techniques that many renowned botanical artists use, taking sometimes three to six months to complete one painting, I personally have the temperament for these less formal drawings. Here is my method for planning this composition.

- I created a preliminary process page of the unusual flowers on this tree.

- I cut a full branch of a tree that had a large fruit and several flowers and leaves.

- I then planned the composition of the cut branch and started to draw the elements.

- After completing most of the elements but leaving some areas unfinished, I decided I wanted to add the notes from my original process page on the reproductive parts and other details.

- I cut open the fruit, drew it, made a smoothie with it (delicious), and scattered some seeds. I finished by adding a flower with cast shadow to bring attention to this unusual flower.

- Last, I added the text and notes to the page.

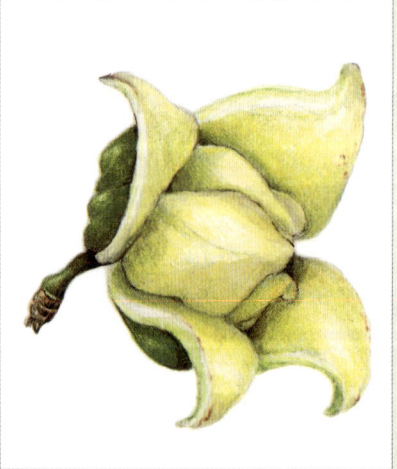 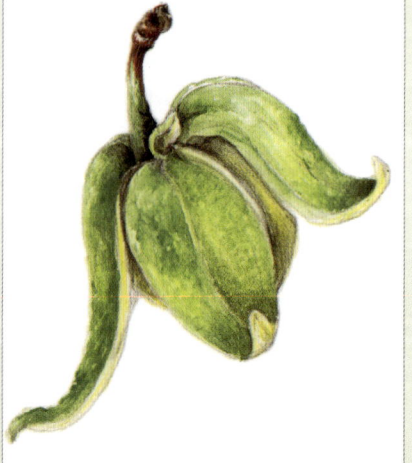

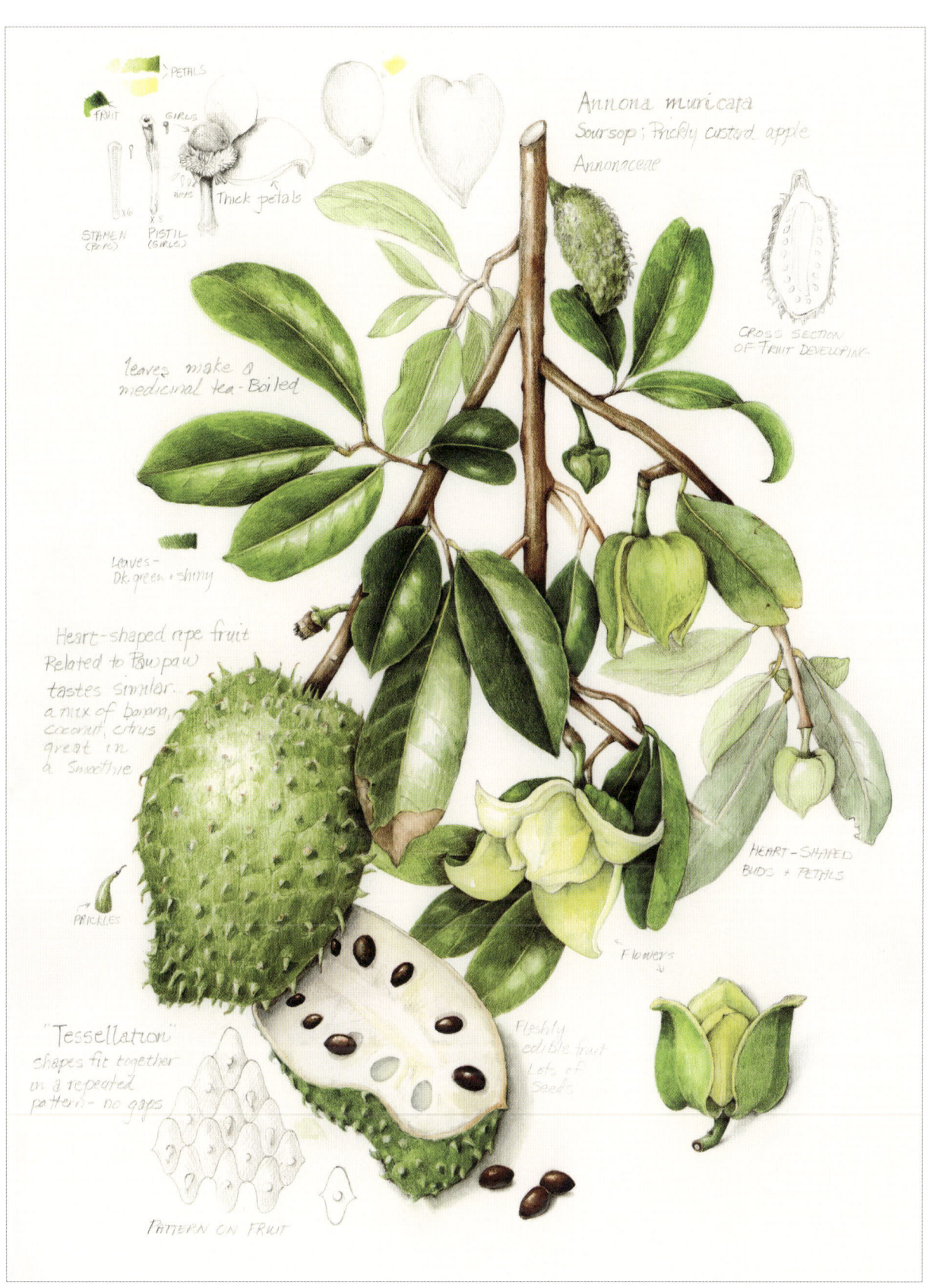

PETALS

FRUIT

GIRLS

STAMEN
(BOYS)

PISTIL
(GIRLS)

Thick petals

Annona muricata
Soursop ; Prickly custard apple

Annonaceae

CROSS SECTION
OF FRUIT DEVELOPING

Leaves make a
medicinal tea - Boiled

Leaves -
Dk green + shiny

Heart-shaped ripe fruit
Related to Pawpaw
tastes similar.
a mix of banana,
coconut, citrus
great in
a Smoothie

PRICKLES

HEART-SHAPED
BUDS + PETALS

Flowers

"Tessellation"
shapes fit together
in a repeated
pattern - no gaps

Fleshly
edible fruit
Lots of
seeds

PATTERN ON FRUIT

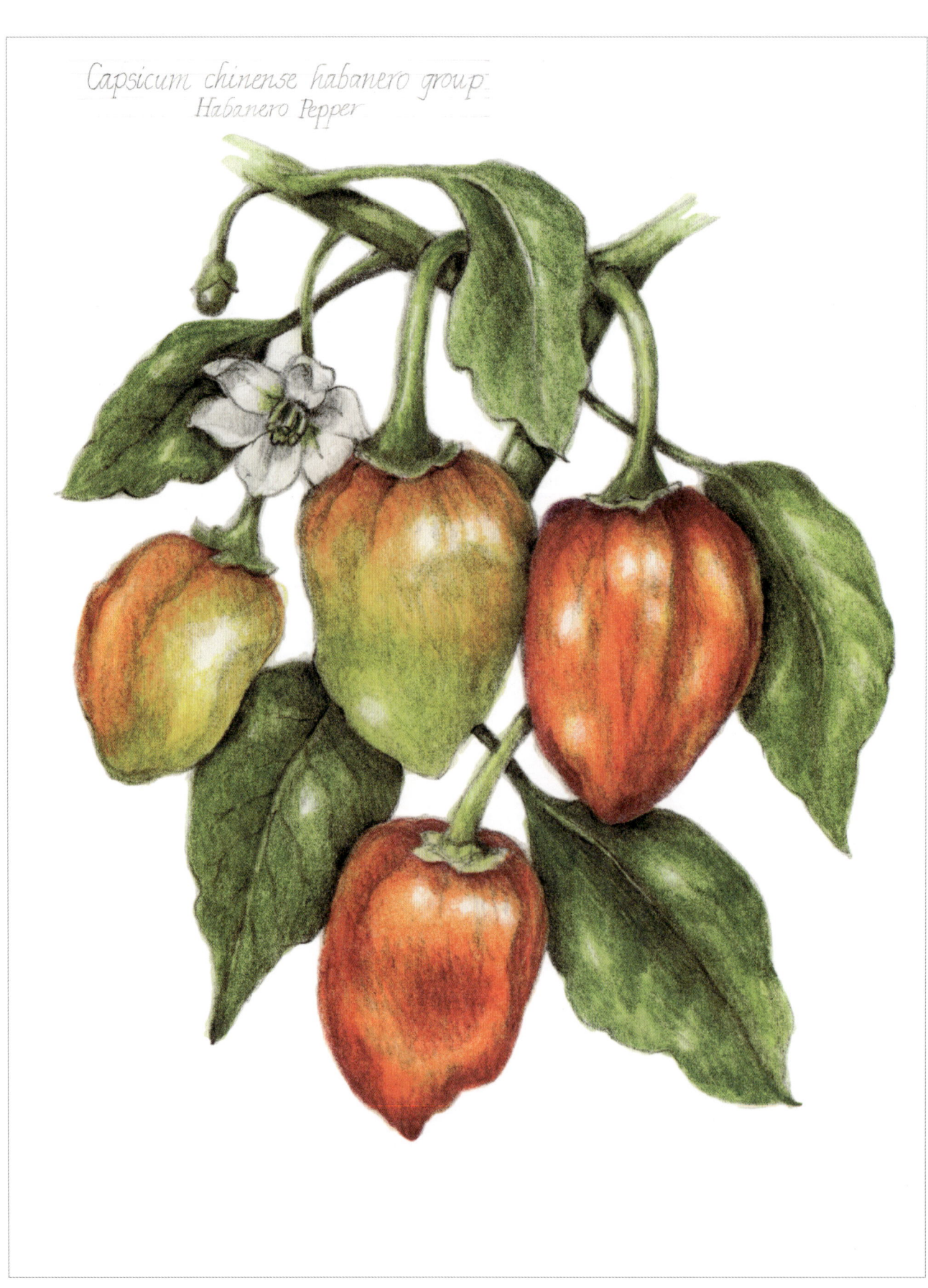

Capsicum chinense habanero group
Habanero Pepper

CONCLUSION: STAYING MOTIVATED

After you read and do the exercises in this book, my hope is that you'll have found your way to your own practice of botanical drawing. Staying motivated may be a challenge, and if so, I encourage you to always be on the lookout for subjects. This is one way to stay inspired and committed to your practice. I'm aware that if I'm feeling stressed, my drawing time is the best way for me to relax. I use drawing as my meditation even if I don't have something specific to draw. Another great idea is to follow along with nature on a daily or weekly basis and keep a sketchbook journal of your plant process pages. You can also follow one single tree all year and document how it changes and develops. Finding a community of like-minded botanical artists on social media can be inspiring, and having friends to draw with is also great. When I was writing this book, Instagram was my favorite way to share and see others' work, but by now, who knows what the venue of choice is for this!

Sources for Subjects

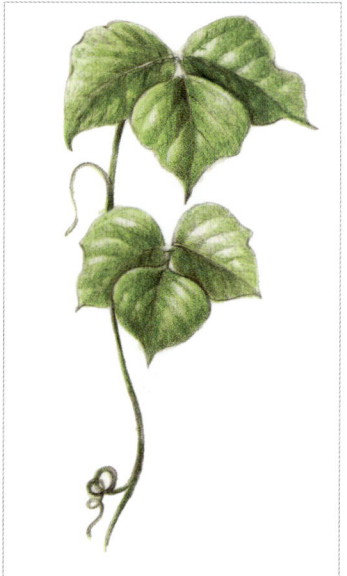

Creating your own relationship with nature drawing will motivate you to keep drawing, because you will constantly find inspiration. Sources for inspiration start in your own neighborhood, including friends' gardens, farmers' markets, florist shops, u-pick farms, your local grocery store, or when you are on a hike in the woods or a stroll in the park. U-pick farms are a great way to observe food growing and are full of wonderful subjects. Having a regular practice of botanical drawing alerts you to all of nature that's around you, whether in a country setting or in an urban area. You'll be tuned in to nature all the time and notice things you'd like to study and draw. Pick these things up, and keep them handy to draw when you can. I keep things for many, many years, and it's remarkable how well nature preserves itself.

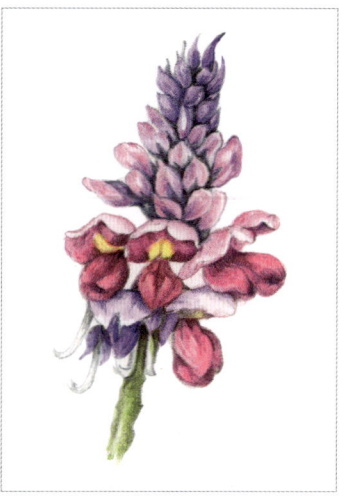

Rare and endangered plants in the wild should not be picked. In a nature preserve or botanical garden, get permission to pick anything. Things that have fallen on the ground are, in my opinion, up for grabs unless noted with a sign. Whenever possible, ask before taking something, and never cut flowers from a planted garden, either in a public park or someone's yard, without permission. Explain why you want the flower. People are often delighted to offer you samples, especially if you show them your drawings! Suggestion: If you have friends who are gardeners, the best gift you can give them is a drawing of something from their garden. When gathering, be careful of poisonous plants such as poison ivy, poison oak, and flesh-eating philodendrons. Pick wild mushrooms for drawing, but please don't eat anything unless you are absolutely sure it is safe.

Evaluate the Components of Your Own Drawing

Dealing with feelings of frustration presents one of the greatest challenges to staying motivated. This often happens when you reach a point in a drawing at which you come up against:

1. Not knowing what to do next.

2. Not knowing if the drawing is finished.

3. Knowing something is wrong, but not knowing what to do about it.

This is the perfect time to start to evaluate your drawing. Use the following checklist to help you figure out where to go next with the drawing. If you evaluate your drawing this way, you'll be able to decide the answers to the three problems above. Give it a try, and see if this self-evaluation process will help you overcome these frustrations that we all feel from time to time.

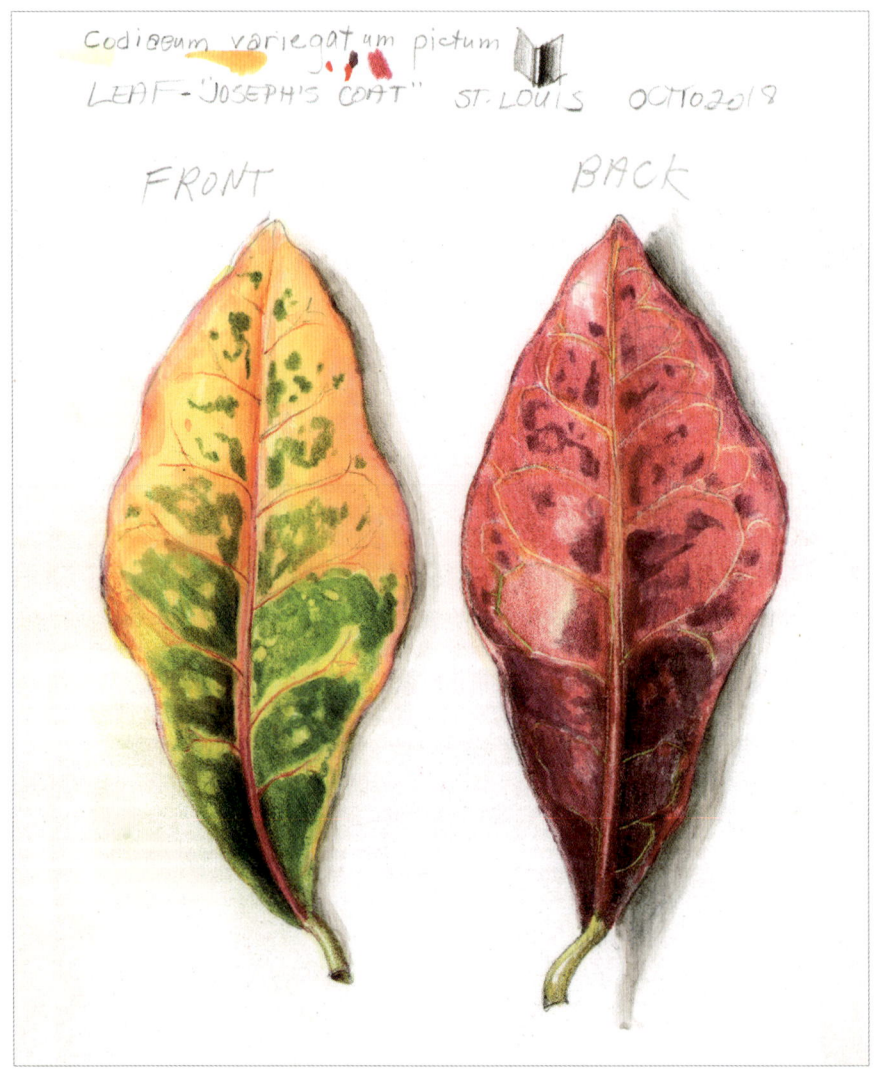

EVALUATE-YOUR-DRAWING CHECKLIST

❏ **SIZE:** Is the drawing accurate in size, proportion, and perspective?

❏ **STRUCTURE:** Does the drawing show the accurate structure of the subject?

❏ **OVERLAPPING AREAS:** Are overlapping areas of the subject clear? Is it obvious what is in front and what is behind? Is there a defining dark shadow underneath or behind the subject in the foreground?

❏ **LIGHT SOURCE:** Is the light source that illuminates the front of the subject appearing to come from the left or right?

❏ **HIGHLIGHT:** Is the highlight in the correct location? Is there a shimmer to the highlight, not just an empty space?

❏ **TONES:** Is there a complete range of nine tones from light to dark? Do the tones graduate and blend seamlessly on the form?

❏ **COLOR:** Are the colors accurate? Is there a complete range of nine tones maintained in color?

❏ **REFLECTIVE HIGHLIGHT:** Is the reflective highlight believable, not appearing as empty space?

❏ **CAST SHADOW:** Does the cast shadow recede and push the subject forward in space? Does the shadow have soft edges? Do the shadow's tones graduate from very dark to very light so that the lightest tone fades into the paper?

❏ **DETAILS:** Are the drawing's details believable and not overshadowing the subject?

❏ **EDGES AND CONTOUR:** Are the edges of the subject sharp without creating a dark outline? Does the drawing appear to be "in focus"? Do your subjects have subtle curves, rather than sharp angles, so that the contour of the drawing is graceful?

❏ **COMPOSITION:** Are the layout, placement, and location of the subject visually comfortable on the page? Is there a clear focal point?

PRACTICE THROUGH THE SEASONS

Here are some things to look for and practice drawing in each season.

WINTER: Dried nuts, curling leaves, branches, and bark.

SPRING: Emerging flowers and leaves, and lots of wildflowers.

SUMMER: Flowers, fruits, and vegetables. Add in a bee or butterfly!

FALL: Colorful leaves, gourds, squash, and root vegetables.

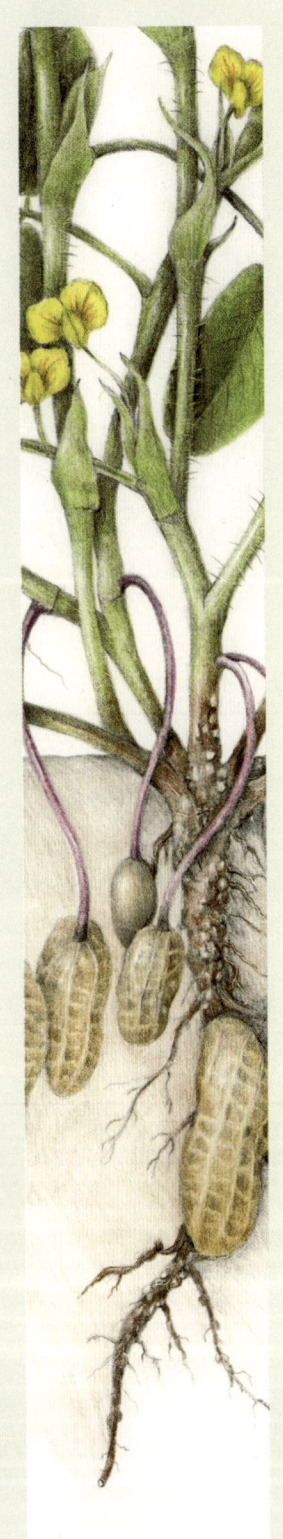

SUMMARY STEPS FOR BOTANICAL DRAWING

Now that you have been following my step-by-step lessons throughout this book, you may notice that I vary my approach depending on my subject, or depending on which way the wind was blowing on a particular day! The important takeaway is to understand the thought process behind these techniques and figure out how you will include them in your own process. There's room for lots of flexibility. You can choose to vary the order of some of these steps and decide to add watercolor washes before grisaille toning, or to leave some areas of a drawing unfinished. The natural world is now yours to play with and enjoy. Whether drawing becomes your daily meditation practice, your weekly journal entry, or your series of botanical masterpieces, make these steps your own and soon you will just naturally know what to do and won't have to think about it. But here's a quick reference that may become a mantra that plays in your head as you work. Many of my students say that they hear my voice in their head as they draw, guiding them along, helping them remember to use a good light source or, most important, *to sharpen their pencil!*

1. Choose your subject first and always draw from a real specimen. Don't draw from my artwork or photographs.

2. With a graphite pencil, lightly draw your form life-size. Start with a center axis line, measure with a clear ruler as needed, and then put your graphite pencil *down*.

3. Analyze your light source—upper left or upper right lighting? Draw a small thumbnail sketch of the geometric form closest to your subject with this correct light-source model. Decide on the placement of the highlight, and outline it on your drawing with a graphite pencil very lightly so that you can erase the placement lines later. Do an optional cross-contour thumbnail sketch to feel how the form bends.

4. Remember to always work with a sharp colored pencil. Never tone with your graphite pencil. Using a dark, neutral colored pencil influenced by the local color of your subject, start to add grisaille toning to your form by first defining the overlapping areas. (For dark subjects, Dark Sepia, Red Violet, or Chrome Oxide Green are good options. For lighter subjects, Earth Green is a good choice.) Do this by lightly toning behind the form in front.

5. Continue to add grisaille toning with colored pencil. Tone light to dark to define the three-dimensional surface of the form with a consistent light source. Build your tones slowly, making sure to achieve a complete range of tones from light to dark. The lightest tone is the color of the paper. Remember, in the beginning, rendering three-dimensional form and structure with correct tones is more important than the local color of your subject. You don't need to completely tone your form, just do enough to define the shadow side and the highlight.

6. Do some color blends to choose the colors needed to describe your subject's local or dominant colors. Create a practice straight or curved tone bar to show the color variations with all of their tones from light to dark.

7. Choose watercolor pencils in the main, or local, color of your subject and mix up a color.

8. Wet your subject with water, give it a few seconds to seep into the paper, and paint a watercolor wash, leaving an empty highlight. Always let the watercolor dry before continuing.

9. Choose colored pencils close to the form's dominant color and layer color on top of the grisaille toning. Keep working between colors and darker tones.

10. Consider adding a reflective highlight if desired, and start to burnish your drawing. Any highlights should be left the color of the paper and later can be developed with lighter tones and colors, filled in as needed so that the highlight doesn't look like an empty spot but instead a shimmering light.

11. With Verithin pencils, sharpen edges and fine lines in your drawing, and add details. With a Dark Sepia colored pencil, darken where contrast is needed.

12. Step back and look at your work, evaluate your drawing following our checklist (see page 173), work on it a bit more, and then give yourself a round of applause!

Thanks for joining me on this journey. Keep up your practice and send me your drawings! Post them at drawbotanical.com/joyofbotanicaldrawing or on social media with the hashtag #joyofbotanicaldrawing so we can all enjoy seeing each other's work!

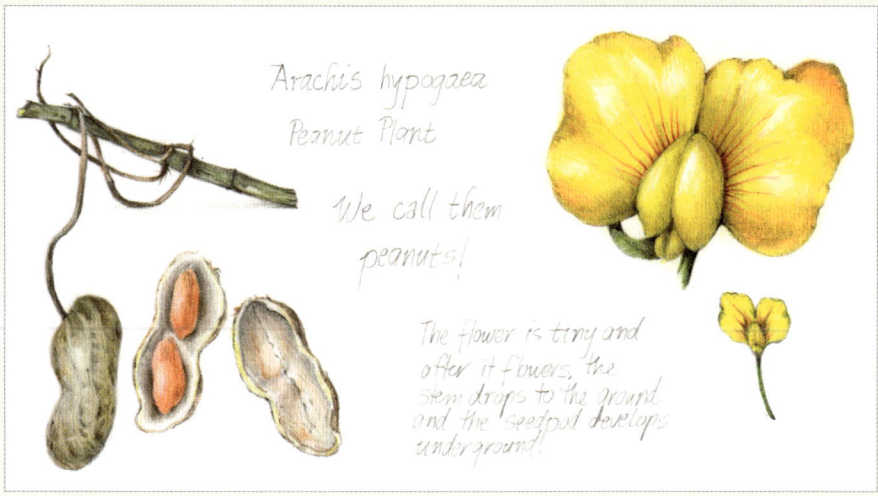

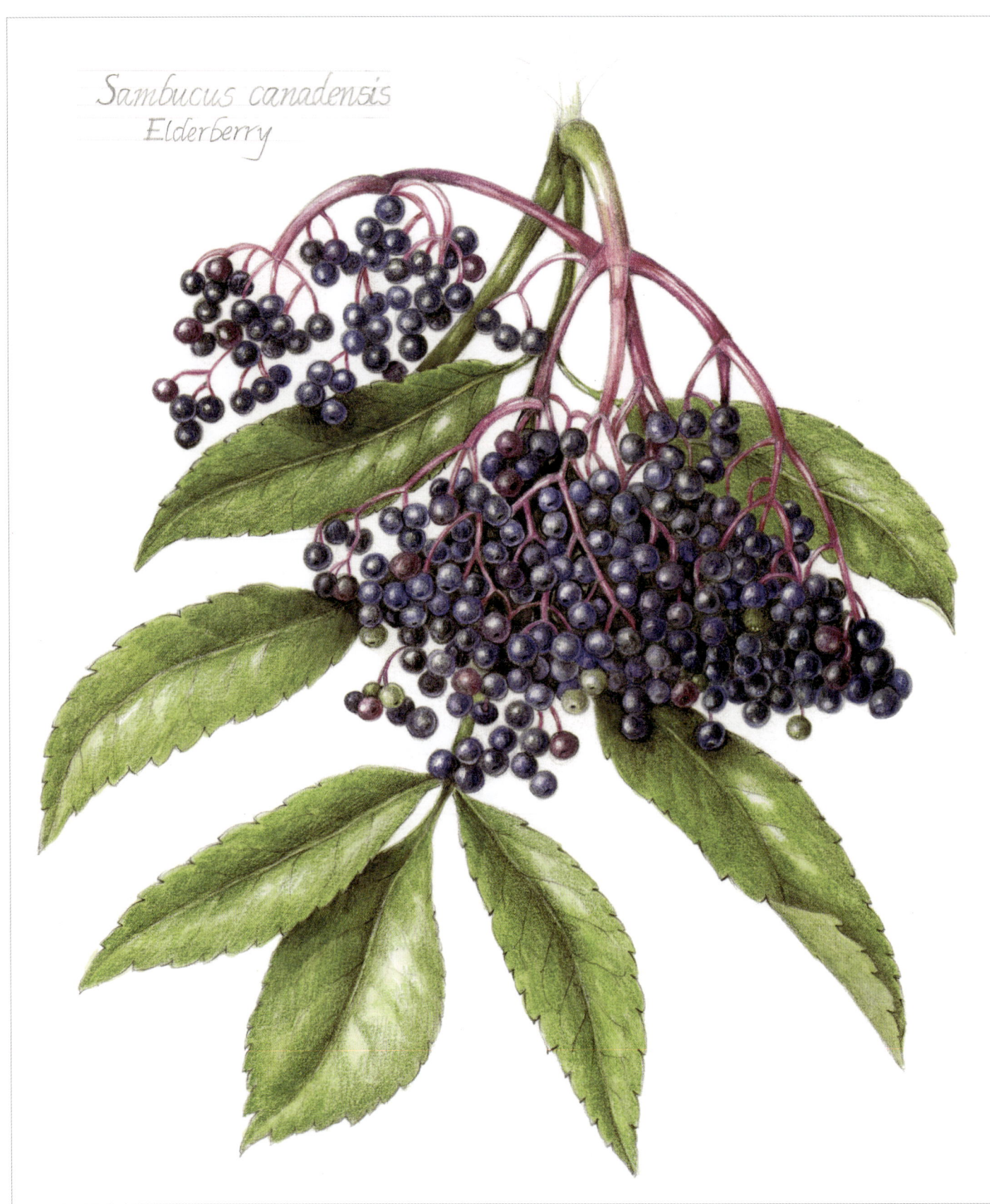

Sambucus canadensis
Elderberry

ACKNOWLEDGMENTS

I want to thank so many people for giving me support and encouragement on my journey with botanical art. I especially want to thank my enthusiastic students from all over the world, for expressing their appreciation for what I teach and for being loyal followers for many years. It is music to my ears when I hear them say "Learning botanical drawing has been life-changing." Some even compare it to falling in love!

Veronica Fannin became my assistant a few years ago, at first helping with graphic design. I now consider her a respected colleague as she has developed into a talented botanical artist and instructor, helping me perfect curriculum and teach workshops around the world and online. Without her help I wouldn't have had the time to create this book, and her suggestions and organizational skills expedited the process.

I want to thank Dina Falconi, herbalist and author of *Foraging & Feasting: A Field Guide and Wild Food Cookbook*, for broadening my knowledge of plants and sharing her undying enthusiasm for nature.

A shout-out goes to the group of talented international botanical artists who join me each year at the National Tropical Botanical Garden. We share techniques while developing a Florilegium collection of art for this spectacular garden. Working together alongside the knowledgeable botanists and horticulturists helps us grow as botanical artists and enhances our artwork and its usefulness in today's world.

My two children designed and created the gardens at my home, which is a daily source of inspiration. It means a lot that they and their growing families share my passion for plants. I look forward to discovering nature all over again with my grandchildren as we walk around the gardens together exploring—planting peach pits and other seeds and experiencing nature's life cycle and magic together.

Thanks to the talented group at Ten Speed Press: my editors, Ashley Pierce and Sigi Nacson, and designer Chloe Rawlins, for enthusiastically supporting this project.

More than anything I want to thank the plants themselves for inspiring and nourishing me!

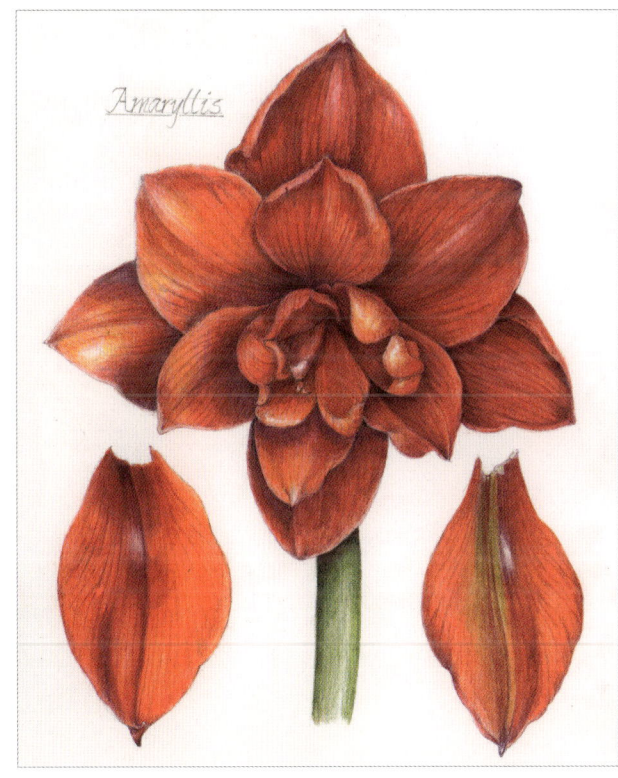

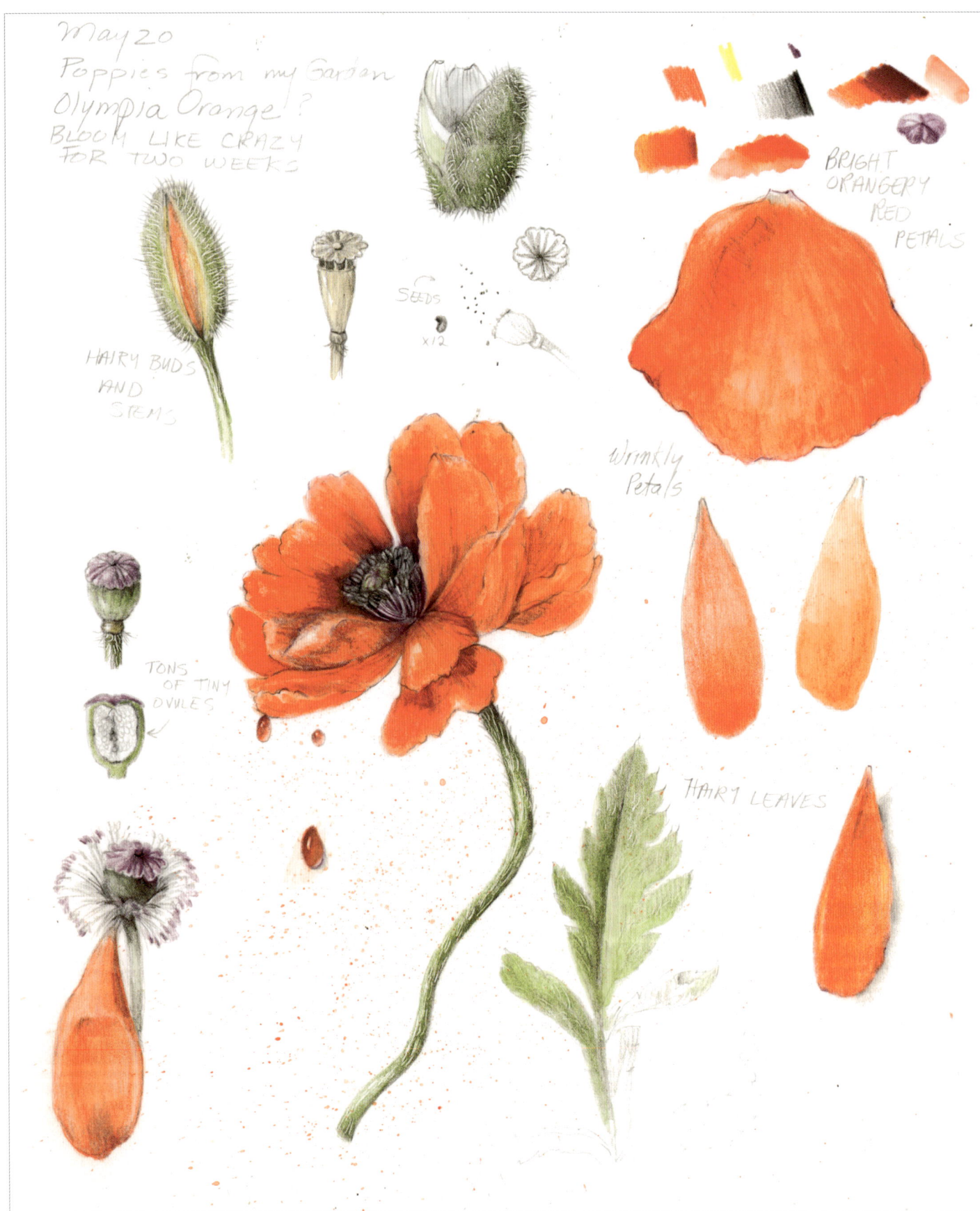

May 20
Poppies from my Garden
Olympia Orange!?
BLOOM LIKE CRAZY
FOR TWO WEEKS

HAIRY BUDS
AND
STEMS

SEEDS
x12

BRIGHT
ORANGERY
RED
PETALS

Wrinkly
Petals

TONS
OF TINY
OVULES

HAIRY LEAVES

ABOUT THE AUTHOR

WENDY HOLLENDER is a botanical artist, illustrator, instructor, and author. She is a leading expert in using colored pencils and watercolor pencils to create detailed botanical drawings and paintings. Her illustrations have been published in the *New York Times*; *O, The Oprah Magazine*; *Real Simple*; *Good Housekeeping*; and *Martha Stewart Living*. Her work has been exhibited by the Royal Botanic Gardens, Kew, the Smithsonian National Museum of Natural History, and the United States Botanic Garden. She is currently an instructor at the New York Botanical Garden, and leads workshops at her farm in Accord, New York; the National Tropical Botanical Garden in Kauai, Hawaii; and in Greece. You can view her online course and publication, *The Practice of Botanical Drawing*, at drawbotanical.com. Wendy loves going to new and botanically inspired places to teach, so invite her to yours! She is the author of *Botanical Drawing in Color* and the self-published *Botanical Drawing: A Beginner's Guide*, and the illustrator of *Foraging & Feasting: A Field Guide and Wild Food Cookbook*.

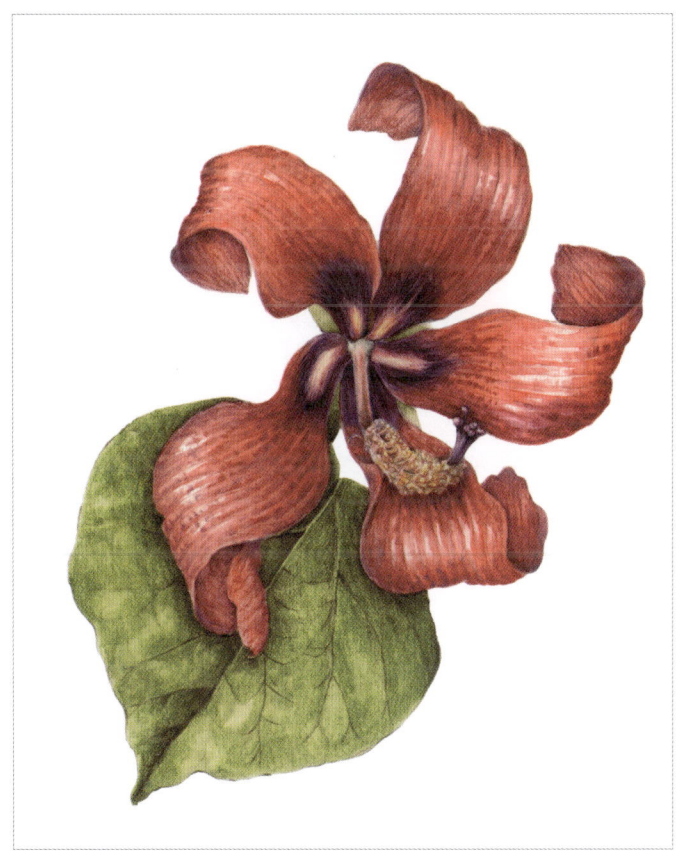

LIST OF ILLUSTRATIONS

Trumpet Vine   Campis radicans

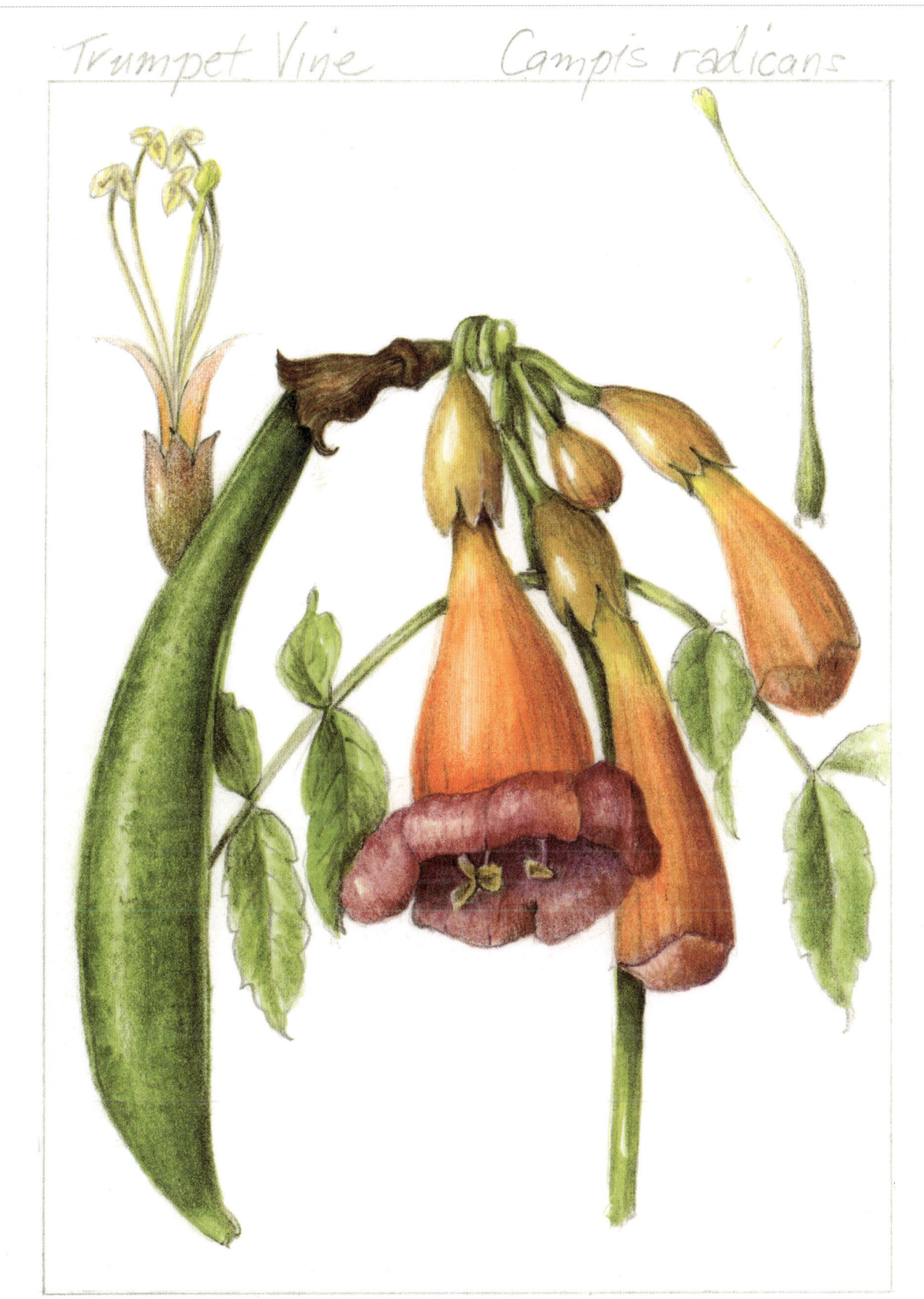

INDEX

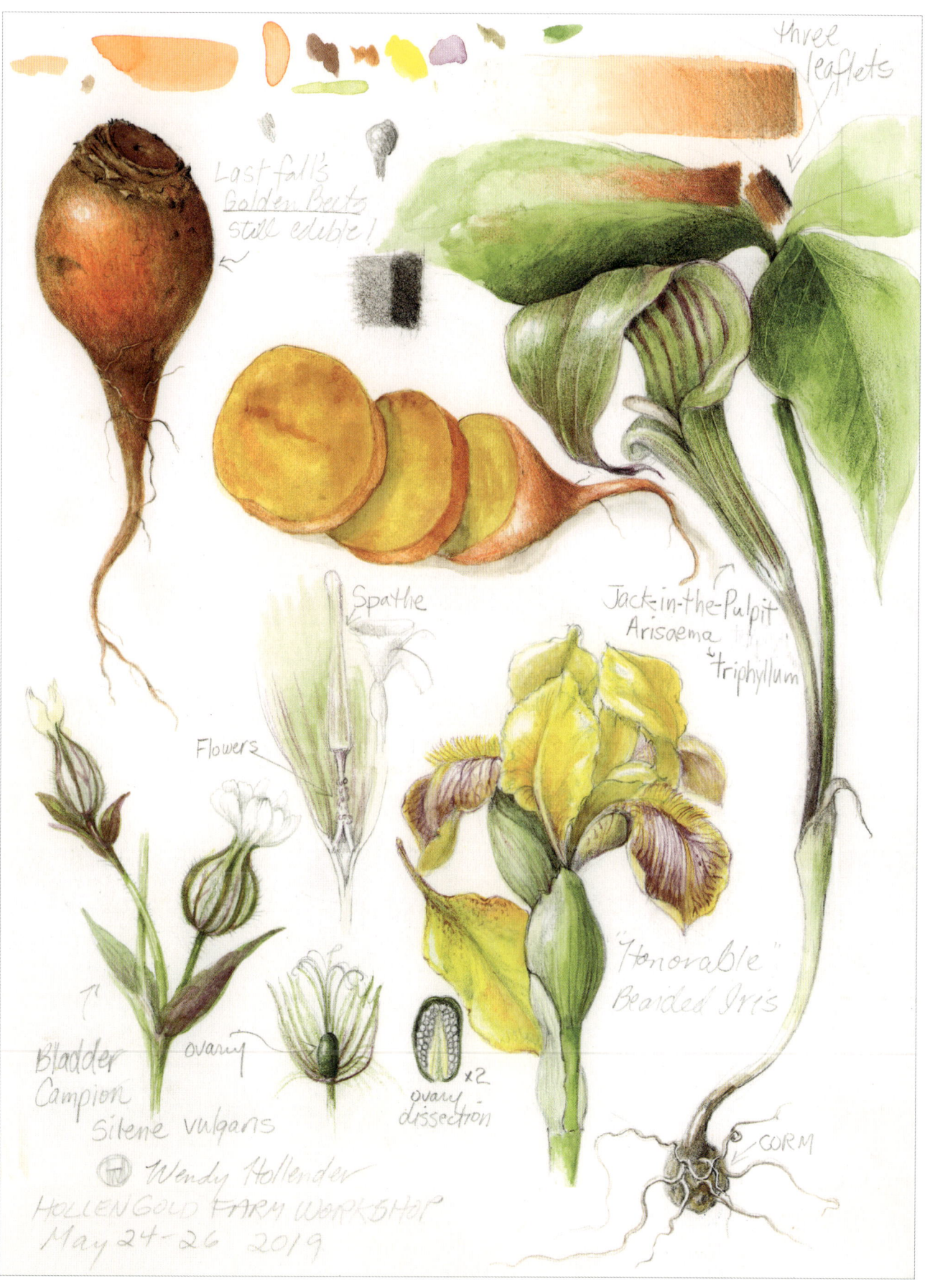

three
leaflets

Last fall's
Golden Beets
still edible!

Jack-in-the-Pulpit
Arisaema
triphyllum

Spathe

Flowers

Bladder
Campion
Silene vulgaris

ovary

ovary
dissection

x2

"Honorable"
Bearded Iris

CORM

Wendy Hollender
HOLLENGOLD FARM WORKSHOP
May 24-26 2019

Library of Congress Cataloging-in-Publication Data
Names: Hollender, Wendy, author.
Title: The practice of botanical drawing / by Wendy Hollender.
Description: First edition. | Emeryville : Ten Speed Press, 2020. |
    Includes bibliographical references and index.
Identifiers: LCCN 2019029973
Subjects: LCSH: Botanical illustration—Technique.
Classification: LCC QK98.24 .H6476 2020 | DDC 581—dc23
    LC record available at https://lccn.loc.gov/2019029973

Trade Paperback ISBN: 978-1-9848-5671-5
eBook ISBN: 978-1-9848-5672-2

Printed in China

Design by Chloe Rawlins

10 9 8

First Edition

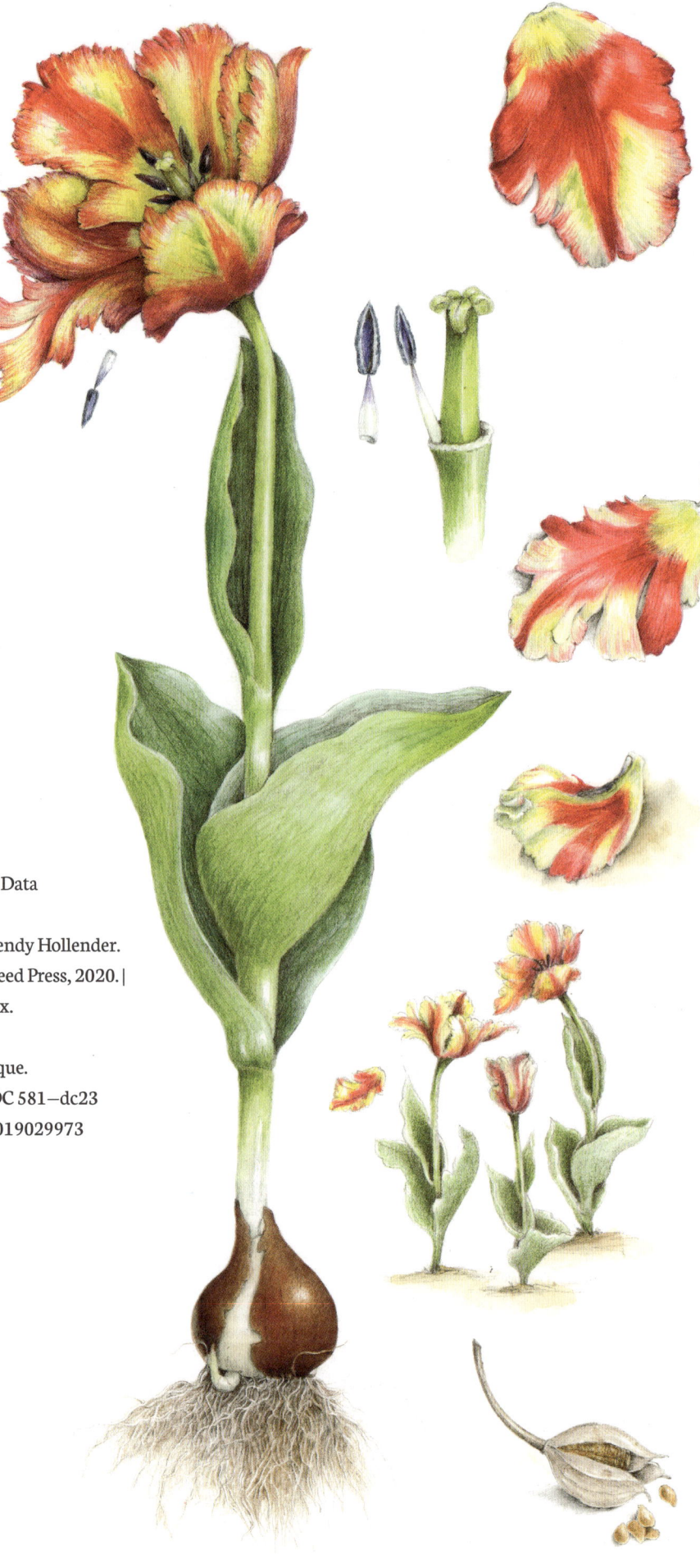